MW00855739

Praise for
Air Born

Jan Davis is a crackerjack astronaut, engineer, and pilot. Combining her father's WWII story of courage aloft with her own soaring career flying the Space Shuttle, *Air Born* excites and inspires.

—Tom Jones, veteran astronaut, B-52 pilot, and author of Space Shuttle Stories

In the tradition of *Flags of Our Fathers*, this is the heartwarming and stirring story of a daughter who became an astronaut and her bomber pilot and prisoner of war father, lost to her for decades but found at last through her diligent and tenacious research. Their brave adventures in the sky and beyond, separated by time, distance, and fate, were intertwined in ways neither knew until the entire story was known. The result is a book with tales of deadly war and dangerous spaceflight that is also suffused with the poignant search of a child, now a successful adult, for a father whom she always loved but who remained a distant figure throughout his life.

—Homer Hickam, author of Rocket Boys *and* October Sky

Air Born is a story of a family legacy and also one of incredible technological advancement. In 1943, when Benjamin Smotherman piloted his YB-40 across the Atlantic in WWII, he could barely have dreamed that the next generation would be exploring the frontiers of outer space. Dr. Davis expertly weaves her own story of training with that of her father's service in the 8th Air Force. The result is a story that is personal and historic.

—Nathan Huegen, Director of Educational Travel at The National WWII Museum

The oft-used word "unique" rarely fulfills its definition. Dr. Jan Davis's book, *Air Born,* not only meets this high bar—it exceeds it in myriad ways. The depiction of her WWII B-17 pilot father's experiences includes his being shot down in 1943 followed by his internment as a POW in Stalag Luft III and near-death slog on the Nazi forced march of prisoners west to Moosburg in the winter of 1945. Following in his iconic footsteps, Dr. Davis combined a mechanical engineering education with an early penchant for flying that resulted in a career as a NASA astronaut that included three missions on the Space Shuttle in the 1990s. Dr.

Davis weaves this captivating story of these two lives knitted together by common vocations and familial unity even though separated by five decades of aeronautical advancements. In so doing, she combines the emotions of these parallel lives with the details of flying two genres of flight, each at the height of their respective time and service, that has kept our nation free for over eight decades. *Air Born* will teach fortunate readers while letting them feel the process in a fulfilling and intimate way. Do yourself a favor—read *Air Born*! It will not disappoint!

> *—J. Ross Greene, author of* A Fortress and a Legacy, *founder and curator of www.WorldWar2Collection.com, and member of the National Museum of the Mighty Eighth Air Force board of trustees*

He flew through the blue battlefield in angry, war-torn skies over Germany in World War II. Decades later, she flew into the expanse of the blue frontier, full of wonder and discovery. He flew the experimental YB-40 aircraft, a B-17 variation that few would ever fly, while she was one of the few to fly on America's pride, the NASA Space Shuttle.

Pilot Ben Smotherman's 1943 view downward from the cockpit revealed the Netherlands, a country in crisis pinpointed for destruction, as his bombardier squinted through the small eyepiece of the secret Norden bombsight, in its day a technological wonder, while her view of the heavens revealed all the vast beauty of the cosmic universe through the sophisticated Hubble Telescope.

Lt. Smotherman's B-17 had limited range and speed carrying its five-hundred-pound bombs, and with no pressurization or heat, its crew endured -50 degrees while aloft. Jan Davis, astronaut and private pilot, and her crew sat atop the equivalent of a bomb and flew faster, farther, and immensely higher with many of the comforts of home. But both experienced the same emotions, one in wartime, and one in peacetime—excitation, apprehension, and fear that did not deter either of them. Father and daughter were kindred spirits whose fascination with flight led them on their life's journey in a close eternal bond.

Jan Davis's quest to discover her father's WWII experiences took three years as she traced his footsteps while referencing his Wartime Log to guide her. She learned about his experiences as a prisoner of war at Stalag Luft III and Stalag VIIA and understood the full impact of what he endured.

Ben Smotherman lived long enough to know that his daughter was an astronaut, but he died just before her first space flight. She was able to fly over to see him in the hospital, affording him the chance to see her in her NASA flight suit. But bonds never break, and through her extensive efforts and determination, his legacy lives on.

—Marilyn Jeffers Walton, author, WWII historian, researcher, and script consultant for Tom Hanks/Steve Spielberg's Masters of the Air *miniseries*

AIR BORN

Two Generations in Flight

The gripping memoir of a heroic
WWII B-17 pilot POW and his
Space Shuttle astronaut daughter

JAN DAVIS

Ballast Books, LLC
www.ballastbooks.com

Second Edition

ISBN: 978-1-955026-76-5

Cover Design by Tabitha Lahr

Author photo: NASA Photo

Photo cover: USAAF photo colorized
by Rick Foss, rfoss@homesc.com

Printed in Hong Kong

Published by Ballast Books
www.ballastbooks.com

For more information, bulk orders, appearances, or speaking requests,
please email: info@ballastbooks.com

To my family for their love and support.
I am richly blessed.

Table of Contents

Introduction

Five… Four… Three… Two… One… Liftoff! Excitement tempered with a bit of trepidation filled me as I abruptly and violently left the launch pad in the Space Shuttle on my first spaceflight mission. Years earlier, in 1943, my father had the same confluence of emotions as he saw the flare signaling him to take off from East Anglia in his aircraft for his first combat mission in World War II (WWII). Anxious to fly into a forbidding sky, he was consumed by a mixture of exhilaration and foreboding. After many years of training for both of us, our emotions on such unforgettable days were intertwined, with each of us going into the unknown.

Decades after my father's wartime missions, I, too, completed missions. Mine were driven by my acceptance into the astronaut class of 1987 nearly fifty years after my father completed pilot training to fight for our country. But there was another mission to fulfill—a far more personal one.

I realized how many connections my father and I had when I read his journal and his Wartime Log, where he carefully and beautifully recorded his detailed experiences. Reading their aged and delicate pages, I was motivated to tell his story and explain how his life influenced mine—far more than I ever realized.

When I was selected to be an astronaut, the first person I called was my mother. Her excitement and love elicited the response, "Put God first, then your country, and go for it!" I can just imagine that she spoke similar words of encouragement when she sent my father off to war, at which point he flew his YB-40 airplane from Florida to England and reported for combat duty in the European Theater of Operations (ETO) in WWII.

Post-war, as his daughter, I lived the American dream provided for me by the sacrifices of not only my father but all those who fought in WWII. I grew up in a good family, was well-educated, had good health, and enjoyed a wonderful career. However, it was not the ideal story as portrayed on television with everything perfect and happy. In fact, my family did not fit the norm of an American dream.

My mother and father were raised in Texas during the Depression and met at North Texas Agricultural College (NTAC). They were married at an early age the day before the Japanese attack on Pearl Harbor, which drew America into a world war. After WWII, they divorced when I was four years old. My parents each remarried, and I had very little contact with my father after I moved to Florida in the second grade. "Daddy Ben" was a career Air Force officer and was transferred all over the country, thus making it more difficult for us to see each other as I was growing up. For as long as I can remember, I have called my father Daddy Ben and will continue calling him that as I tell his story.

Only recently, I have learned about my father as a pilot, a prisoner of war (POW), and a career Air Force officer. With the help of his wartime journal (written to my mother) before he was shot down, his POW Wartime Log, various notes he made, and his Air Force flight records, I have slowly been able to recreate the story of his life so long ago. In addition to those records, I was able to conduct research using archives, books, libraries, museums, websites, and informative Facebook pages.

It was incredible how much information existed—I just had to take the initiative to find it. It was truly an international odyssey that encompassed a wide variety of domestic and international locations. The United States Army Air Forces (USAAF) headquarters and the various B-17 airplane production facilities were in the United States. The WWII bases where many of our boys were sent were in East Anglia, England, the closest they could get to the English Channel and the North Sea to descend upon their targets in Europe. The prison camps were scattered across the Axis powers' wartime acquisitions, though my focus was on the ones in Germany. And my father's plane crash occurred in the Netherlands, necessitating a visit there to begin following in his footsteps.

Throughout my research, everyone in these countries has been most helpful to me. It is heartwarming to find out the interest in this history and the preservation of the stories about the men, machines, airfields, assembly factories, memorabilia, uniforms, and anything associated with the Mighty 8th Air Force and the ETO.

My journey of discovery was very meaningful to me and revealed some of the answers to the unknowns in my own life, long sought after but often elusive. My determination to discover my father's experiences allowed me to put together the puzzle pieces in my own life. The telling of "his story" resulted in the telling of "our story."

Even though I did not have much time with my father, he influenced me more than he knew and inspired me to become a pilot and an astronaut. Combing through my father's records, researching the vintage airplanes and long-forgotten airfields, and examining USAAF pilot training manuals and procedures all shed light on the blanks in his story for me.

There are already many good books about the air war in Europe, and I have not attempted to make this a history book or a reference book about the ETO. I simply wanted to tell my father's story with ample details about the war around him. Hopefully, this biography of my father and my personal memoir as an astronaut will add to the preservation of the history of WWII as well as the space program.

I am grateful to my sister, Darby Smotherman, for sharing the personal effects of our father. She uncovered documents, medals, and records that had been stored away for decades. When we sifted through the old documents to recreate his life, Darby and I could jointly solve the puzzles of the past. Like so many WWII veterans, Daddy Ben never talked about his wartime experiences, including his days held captive by the Germans. Even though Darby was raised by him, and I was not, we pooled our newfound knowledge to paint the final picture of his life.

I am thankful for this renewed relationship with Darby, and that is one of the best byproducts of this book. In his death, he reunited us. I hope the rest of the Smotherman family will appreciate our efforts in years to come.

I would also like to thank my family and friends who supported me on this extensive journey. I especially appreciate the support of my husband,

Dick Richardson, who helped review, critique, and edit this manuscript as well as accompanied me on many of my trips to do research for this book.

Finally, let me take this opportunity to thank those who have served in the United States military to preserve our freedom, including the combatants, instructors, ground crews, support personnel, military leaders, and families. Some made the ultimate sacrifice, and some sacrificed in other ways. I am eternally grateful.

Jan Davis

Chapter One
Texas Born and Bred

In the early 1900s, west Texas was still a frontier, bustling with cowboys, roustabouts, ranches, and hardy pioneers. Amidst that dusty and dry country, Benjamin Franklin Smotherman was born on April 10, 1919, in Fort Worth, Texas. Ben was the sixth child, and the Smotherman family would eventually have three sons and five daughters.

Raised in a strong Christian family by his hardworking parents, Ben learned how to be resourceful and creative at an early age. Boy Scouting was a big part of his life and furthered his passion for arts and crafts.

At some point in his youth, Ben's love of aviation emerged, as shown in Figure 1-1. In addition to that passion, around the age of twelve, he began his lifelong hobby of writing and drawing cartoons. He created comic books titled *Adventures of Bud Tuttle*. He also authored the paper books *The Last Frontier, Castaways, Wolf Dogs*, and *Murder As You Dance*. The cartoons and story lines in the self-bound paper books were very creative and showed what was going on in the mind of a teenager in the 1930s (Figure 1-2). For example, the tales of Bud Tuttle included flying biplanes, cowboys and horses, young love, and animals. The books of cartoons demonstrated Ben's talent and drive. It is a hobby that would serve him well.

Ben's love of aviation continued, and he set about achieving his dream of becoming a pilot. Pursuing a flying career, he enrolled in the Reserve Officers' Training Corps (ROTC) in both junior high and high school. A handsome young man, as shown in Figure 1-3, he graduated from North Fort Worth High School in 1937. After graduation, Ben went to work and saved money to go to school. He always knew that he wanted to be a pilot in the Army Air Corps, so he needed to attend college. As a hard worker who was highly motivated, he ultimately achieved that goal.

Figure 1-2: Page from Adventures of Bud Tuttle, cartoon books drawn by Ben Smotherman, ages thirteen through sixteen

Figure 1-1: Ben Smotherman as a young boy when his love of airplanes began

Figure 1-3: Ben Smotherman as an ROTC student in high school, 1936

In 1939, Ben enrolled at NTAC, now the University of Texas at Arlington but, at that time, part of Texas A&M University, and he lived with his sister, Jack Darby, and her husband. Simultaneously, he was active in ROTC, learning to fly as an aviation cadet, while majoring in industrial aeronautical engineering.

I have Daddy Ben's first flight logbook, which shows his first recorded flight as a student pilot was June 25, 1940, in Grand Prairie, Texas, when he was twenty-one years old. What a thrill it must have been for him to finally be able to take flying lessons! I remember a similar thrill when I started taking flying lessons, but I was twenty-eight years old.

My father did most of his early flight training in a J-3 Piper Cub. Ben's first solo flight was on July 16 after eight hours of flying time. There is nothing like flying on your own when the instructor climbs out of the cockpit and says, "It's all yours!" I know that feeling, and no matter how many hours of flying you accumulate, that first solo flight is the most exhilarating. Ben's first solo cross-country was from Grand Prairie to Dallas to Fort Worth and back to Grand Prairie in a little over an hour. I have been cross-country in a Piper Cub, and in the open cockpit with the wind in my face, it was truly fun.

Ben was finally on his career path as an airman, and he could now take passengers in the airplane. One of his favorite passengers was my mother, Dolly Jo Gantz. The youngest of four children, she'd been born in Comanche, Texas, on October 24, 1924, but raised in Dallas, Texas. The Gantz family had immigrated to Texas from Germany in the mid-1850s, and the name Gantz had been documented then as Gans, Ganz, or Ganss.

My mother graduated from Sunset High School in Dallas in 1940 at the tender age of fifteen. In addition to being an honor student, she loved to dance. She took ballet and ballroom dancing and knew all the popular dances. Dolly Jo entered NTAC as a freshman in 1940. My father was apparently smitten by her! He kept a calendar of each time they met or went on a date. According to his calendar, the first time he saw her was October 25, 1940 (the day after she turned sixteen), and they later met on November 4.

For their first date, my parents went to the NTAC Engineers' Ball on March 1, 1941. In my mother's scrapbook, she noted, "I wore a white net dress and had a wonderful time." There were lots of dances, and one of my mom's girlfriends wrote a note to her, saying, "Boy, you really looked like you were in heaven when you were dancing with him." I can just see my parents dancing their way through the big band and swing era of the 1940s. My father appreciated music and dancing as well.

In April 1941, he wrote his first song to my mother:

I'm falling in love with the girl of my dreams.
The girl of my dreams—that's you!
You came to me out of a world dark and dreary.
I'm glad and happy—it's true.
Though dreamers dream that dreams come true,
I dream of one, and that one's you.
I'm falling in love with the girl of my dreams.
Dolly, my darling, it's you.

What a romantic! I always thought that Daddy Ben wrote well, and I imagine that he could sing well too. I did not know he wrote songs for my mother until I uncovered these gems.

On a day in May, an article in the school newspaper said, "Ben Smotherman took Dolly Jo Gantz flying Friday. What we're wondering about is why all his classmates laughed when he announced that he was going to practice landing with the automatic pilot." Another snippet from the May newspaper: "Ben Smotherman took Dolly J. Gantz flying Saturday morning. They went to Dallas and flew over Dolly's home." My mother has very fond memories of these flying "dates" in the open-air cockpits, and she told me all about them.

Ben Smotherman was elected president of his sophomore class, the class of 1941, at NTAC. In the yearbook, by his good-looking class officer picture (Figure 1-4), it noted, "Ben Smotherman, smiling prexy of the class of 1941, is a day-dodging second lieutenant from Fort Worth…Pipe smoking and painting are his favorite means of diversion…Member of Engineering Society…Officers' Club. Ben, who was a royal escort in the Coronation, wants to be an aeronautical engineer."

My father's leadership abilities and popularity were evident in those early years at NTAC. Handsome and an extremely talented artist and engineer, he was obviously a very interesting person at that time, as he was throughout his life. My mother and father continued to date off and on through May and the first week of June. Then, they went to a dance together on July 11 and became engaged on July 29, 1941! As my parents' love was burgeoning, so was the war in Europe. As a result, the USAAF was created on June 20, 1941, to provide unity of command over the Air Corps and the Air Force Combat Command. Major General H.H. "Hap" Arnold was designated the commander of the USAAF.

Figure 1-4: Ben Smotherman as president of his 1941 class at North Texas Agricultural College

After Ben's graduation from NTAC, he was enlisted in the Army Air Corps Reserve as a second lieutenant and aviation cadet on June 9, 1941, **shortly before the Army Air Corps was folded into the USAAF. He was a second lieutenant and an aviation cadet. Before he was called to begin training, Ben worked as a draftsman at The Austin Company in Fort Worth throughout 1941.**

On December 6, 1941, my parents were married in Love County, Oklahoma. In Mom's scrapbook, she saved two tickets from her high school's football game that occurred on December 6. A note beside the tickets revealed her deep feelings: "One football game we never saw." They apparently eloped to Oklahoma that day, and I'm not sure they told anyone. Mom was only seventeen! From my father's journal for my mother that he kept during the war, he said of the wedding, "Our glorious day, for it is the day we shall always remember—our wedding day! You were so sweet in your gray suit. We met in the Dallas bus terminal at 6 a.m. that day and rode to Marietta, Oklahoma, for our simple ceremony. We were so happy that day and the days following."

The next day, the Japanese bombed Pearl Harbor, Hawaii, and that date marked the beginning of WWII for the United States. President Roosevelt called December 7 "a date that will live in infamy." We were at war, and the lives of my newlywed parents would be changed forever.

Pilot training for aviation cadets consisted of four flying phases and averaged about seven to nine months total, depending on availability of planes, instructors, and slots at training facilities. To help streamline the program, initial military training, as well as administration of the classification of officers, was moved into a preflight program.

To accommodate the new preflight training program, the War Department established three Replacement Training Centers in January 1941 at existing training facilities—Maxwell Field in Alabama, Kelly Field in Texas, and Moffett Field in California. The centers were for classification and preflight training of candidates for pilot, bombardier, and navigator. In the initial classification stage, it was determined which position cadets were best suited. Then, they all progressed to replacement training.

My father attended his replacement training at Kelly Field in San Antonio, Texas, beginning January 7, 1942. His journal reflected his thoughts: "On January 6, one month later [after they married], I kissed you and left for San Antonio to begin my training in the Air Corps. You came to see me there, and it was so good to be together again."

My father went through five weeks of training in San Antonio. According to his aviation training books from NTAC, he had already studied aerodynamics, navigation, and engines. Therefore, he had more training than most of his replacement center classmates. The lack of instruction in basic technical and aviation knowledge due to the short training time was a hindrance to most of the students.

After replacement training or preflight training, the officers went to their respective training schools. Those who successfully completed their preliminary training and became classified as pilots were sent to flight training.

The first phase of flight training was called primary training, which was when aviation cadets learned to fly in a small two-seater aircraft. The secondary phase was called basic training, during which pilots transitioned to a heavier plane with more complex controls. In advanced training, they learned to fly a still more powerful machine with controls similar to combat aircraft. Finally, there was transition training, where they learned to fly the airplane to which they would be assigned for combat, whether single engine or multi-engine.

The USAAF pilot training program in the early 1940s was accelerated at such an incredible rate that civilian contractors were selected to operate a large number of newly established primary flying schools. The basic and advanced flying schools were operated by the USAAF. Ten civilian contractors started operating the primary training schools in 1939.

Training aircraft used in the primary flight schools were mostly Fairchild PT-19s, PT-17 Stearmans, and Ryan PT-22s, although a wide variety of other types could be found at the airfields. As my father started flying PT-19s in early 1942 at Garner Field at Uvalde, Texas, it was clear that he had an advantage due to his pilot and aviation training while he was at NTAC.

Ben's journal indicated that my mother visited him there. Mom worked as a U.S. government civil servant secretary, so I imagine that she was able to travel to visit during weekends and, of course, graduation.

The PT-19 was a tandem two-seat cantilever low-wing airplane that I am sure was a joy to fly. Flying at speeds around one hundred miles per hour, the aviation cadets learned quickly to respect the wooden and canvas open-cockpit airplane. I can just imagine that, when my father was in training, he realized his dream was coming true. A picture of him with his perpetual smile while he was in primary flight training is shown in Figure 1-5, reflecting that inner joy.

In addition to experiencing the fun of learning to fly a new aircraft, he also enjoyed the camaraderie and shared goals of his fellow classmates. That is exactly the way I felt when I started my astronaut training—I was on my way to achieving my dream. Furthermore, the fellowship and mutual interests of my fourteen other like-minded classmates in the astronaut class of 1987 bonded us for life.

Figure 1-5: Ben Smotherman at Primary Flight Training, Uvalde, Texas, 1942

On February 28, 1942, my mother gave a Bible to my father. In it, she inscribed, "Ben dearest, I request that you carry this book with you when you are in flight or in danger of any kind. Accept it with all my love and prayers. Dolly." She probably presented him with the Bible, a memento from her, when they were saying their goodbyes while he was in flight training.

The memento gifted so long ago is significant to me, as I obtained that Bible from my mother and carried it with me whenever I flew as a pilot. As an astronaut, I wore royal blue flight suits whenever I flew the T-38 jets, and the Bible fit nicely into one of the leg pockets. I also carried the Bible with me on one of my Space Shuttle missions, sealed in pink plastic wrap and stowed in one of the lockers on the Space Shuttle.

My father's basic flight training occurred at the newly opened Waco Army Airfield in Waco, Texas, from May 4, 1942, to July 3, 1942. He had already accumulated sixty hours of recorded flight time in the USAAF from his primary flying school when he arrived at Waco.

Fortunately, he was able to visit my mother in Dallas and his family in Fort Worth on the weekends. Meanwhile, my mother continued to go to school and work, and it was hard on their young marriage to be separated for so long. Such was the case for many men and women entering the military services to go to war. Families sacrificed, as did the trainees, and the whole country felt the impact of the conflict.

In addition to operating a single-wing airplane of greater weight, horsepower, and speed, including the BT-9, BT-13, BT-14, and BT-15, Ben was taught how to fly in formation and on cross-country flights. He was also instructed on night flying and instrument flying (instrument flight rules, or IFR). There were other new procedures to learn as well, such as the operation of a two-way radio and a variable pitch propellor. During my own training as an instrument pilot, I also progressively learned how to do these things, including "flying under the hood" to learn how to be totally dependent on following my aircraft instruments.

My father conducted his advanced training in Lake Charles, Louisiana, at Lake Charles Air Force Station (AFS). Daddy Ben's flight records indicated that he flew the AT-6 aircraft, nicknamed "The Texan," from July 7 to September 6, 1942. The Texan was a more complex airplane than anything he had flown before, as it had retractable landing gear, a larger engine, a variable-pitch propellor, and hydraulics.

My mother visited him in Lake Charles occasionally, as evidenced by the postcard and memorabilia that she had in her scrapbook. She wrote by the saved postcard, "Dear Lake Charles and its mammoth mosquitoes." A picture of my mother and father on his graduation day, September 6, 1942, is one of my favorites of them and is shown in Figure 1-6. My mother's caption by this photograph in her scrapbook was, "We were so happy."

Along with receiving his silver wings after completing advanced pilot training, Daddy Ben became a second lieutenant in the USAAF. At that point, he was no longer an aviation cadet or in the reserve—he was an active-duty officer.

As I know from my own flight training, it was just the beginning for my father. Next was transition training, consisting of the upgrade to multi-engine and very powerful airplanes, and the training necessary for combat and working with a crew.

I remember when I received my silver astronaut pin as a symbol of the completion of my basic astronaut training. After that, I was eligible to be assigned to a spaceflight mission. Likewise, my father wore his USAAF wings for the rest of his military career—not only was he then an officer, but he was also a pilot.

Figure 1-6: Ben and Dolly Smotherman on the day he earned his wings at Lake Charles, September 6, 1942

Earning his wings was a huge milestone for my father, but there were many possible roles for him that lay ahead. He did not know what type of aircraft he would fly in the USAAF, where he would train, what he would do, or who his fellow fliers would be. But he and my mother knew that the most probable destiny for him would be to fly in wartime. There was no denying that it was what he had been trained for, and fate wasted no time calling his name.

Chapter Two
Off We Go...

Flight training was occurring at such a fast pace in 1942 that my father received his silver wings on September 6 and was transferred to his next assignment on September 8. He was assigned to Hendricks Field in Sebring, Florida, for transition training and reported on September 14.

It was at Hendricks Field that my father completed his last training phase on the flying school syllabus for an individual pilot. Pilots who graduated from USAAF flight training were assigned to transition training in the type of plane they would fly in combat. For my father, this plane would be the B-17 bomber. His first flight in a B-17 was on September 16, two days after his reporting date at Sebring! The gigantic multi-engine airplane was called the Flying Fortress for a reason—its massive size and sturdy structure. That must have been a phenomenal flight after having only flown single-engine airplanes!

I learned to fly in a single-engine airplane, and I remember the first time I flew at the controls of a twin-engine T-38 jet. When I became an astronaut, we received ground school and flight training in the T-38. I don't know how my experience during my first flight in the T-38 compared with my father's first experience in the B-17, but it was thrilling. With so much extra power and speed, it was a lot more fun!

Ben transitioned to first pilot of the four-engine B-17 heavy aircraft. He and his B-17 crew flew daily missions to bomb sand spits (beach landforms off coasts or lake shores) and simulated submarines, learning how to navigate accurately. All in all, he was in Sebring, Florida, from September 14 to November 18, 1942, and his diploma from Combat Crew Training School specified that he completed the course of instruction for first pilot.

The B-17 Flying Fortress became the third most-produced bomber of all time, with a production of over 12,700 aircraft built for WWII. My father's flights in the B-17s during the transition phase were mostly with the B and E variants. Models A through D were designed for defense operations, but the large-tailed B-17E was the first model primarily focused on carrying its own guns for warfare (Figure 2-1).

Figure 2-1: The B-17E at Hendricks Field
(Hendricks Field yearbook)

There were ten positions on a typical B-17 crew: pilot, copilot, bombardier, navigator, flight engineer/top turret gunner, radio operator, ball turret gunner, two waist gunners, and tail gunner. Part of the combat crew training was to learn the coordination of the various positions on a crew and understand their functions.

Incidentally, I had the opportunity to fly in a B-17G, Texas Raiders, in Huntsville, Alabama, in May 2021. With the start of each engine, it sputtered and huffed and puffed, and I wondered if it would catch. Smoke billowed, and oil drained onto the ground. It was a messy and clunky airplane but beautiful to watch come alive. As each engine roared to life, the plane lurched and coughed until each of the four engines started.

I recall strapping into my seat in the radio room and looking around at the plug for the electrically heated flight suit, at the radios, and at the exposed cables for the flight controls. Then, the squeal of the brakes let me know that we were on our way as we taxied out. I tried to imagine what my father had thought when he flew a B-17, especially when he was flying into combat. It was a thrilling and emotional ride for me, and I enjoyed sticking my head out the window since we were only ten thousand feet above the hills of Huntsville, Alabama. The pilot greased the landing to end a memory that I will always have. There is nothing like flying in a warbird.

After learning to fly the B-17 and graduating from Combat Crew School, Ben was ready for operational unit combat training, designed to teach flight crewmembers how to blend their individual skills into a team and form an effective combat unit.

His total flight time after completing his operational unit combat training was 569 hours, and he was then promoted. First Lieutenant Smotherman was ready to face the enemy in the ETO.

Before he left, he recorded in his journal the frantic days he and my mother spent before his departure: "Then came my first leave—six days—in which we were rushed around trying to see everybody, and we didn't have much time to ourselves. We expected me to leave for overseas then." I'm sure that their vacation was a very special time in my parents' lives, as my mother and father knew that it would not be long before Daddy Ben would be sent to war.

Britain had been anxious for America to join them in fighting the Axis powers. Post-Pearl Harbor, they finally got their wish. In desperate straits, Britain had battled nearly alone as Hitler swallowed up more and more Allied nations. American pilots were being trained as quickly and as safely as possible while aircraft were in production, and the British welcomed them to the quickly constructed air bases that would accommodate them. Britain's war began in September 1939. Royal Air Force (RAF) pilots with their bombers and Spitfires were flying missions at night against German positions. Unfortunately, the night bombing was not very accurate and resulted in carpet-bombing that did little more than destroy the German civilians' morale.

Meanwhile, in the United States, USAAF activated the 8th Air Force at Savannah, Georgia, on January 28, 1942. The 8th Air Force's Bomber Command (BC) was responsible for England and Western Europe.

The 8th Air Force was commanded by Major General Carl A. Spaatz, who transferred its headquarters to England in July 1942. Brigadier General Ira C. Eaker led the Mighty 8th BC. When Spaatz transferred to the Mediterranean Theater in December 1942, Eaker took command of the 8th Air Force. At that time, Eaker had only a handful of B-17s in the European Theater. Soon, more came, along with B-24s that were used by a variety of American bomb groups. At the same time, air bases were being constructed in England at a fast pace to accommodate the Allied airplanes and aircrews.

While the night attacks of the RAF continued, the 8th Air Force planned and executed America's daylight strategic bombing campaign that complemented the British nighttime attacks. Up until then, the RAF had been nearly decimated. Once in place, the 8th Air Force pursued high-altitude daylight precision bombing against specific target systems—aircraft factories, electric power, transportation, and oil supplies—in an attempt to destroy Germany's ability to wage war.

The young men who were in training stateside as USAAF pilots were well aware of the wars going on in the Pacific and in Europe, and most of them were eager to go into combat. Crew training was stepped up, and many men going overseas were replacements for those killed in action (KIA). After flight crews were trained and deemed ready for combat, some, including my father, were processed at Salina, Kansas, at the Smoky Hill Army Air Field.

It is interesting to me that one of my favorite "out-and-back" flights from Houston in T-38 jets with National Aeronautics and Space Administration (NASA) was to Salina, Kansas! Although the airfield was different when I flew to Salina compared to the 1940s, the unique flying connection with my father was still there.

During Daddy Ben's processing for entry into WWII, he found out that he would be part of the YB-40 aircraft detachment at Biggs Field at El Paso, Texas. He was also assigned to the 92nd Bombardment Group (BG), 327th Bomb Squadron (BS), of which he would be a member throughout his time at war.

The 92nd BG was activated at Barksdale Field, near Bossier City, Louisiana, on March 1, 1942, and later moved into operations at MacDill Field, Florida. In Sarasota, the 92nd BG's first commander, Colonel James S. Sutton, became acquainted with V. T. Hamlin, the cartoonist who created *Alley Oop*, a nationally syndicated comic strip. Mr. Hamlin designed squadron insignia for each of the four squadrons in the 92nd BG (325th,

326th, 327th, and 407th), with each insignia showing a different bomber pose. The leather flight jacket of my father's friend, pilot Bill Stewart, proudly shows the squadron patch for their BS, the 327th, in Figure 2-2.

Figure 2-2: Emblem of the 327th Bombardment Squadron, featuring cartoon characters Alley Oop and Dinny

All the proposed YB-40 planes were assigned to the 327th BS. Only twenty-five were built, and only twelve YB-40s made it overseas for combat duty. A list of YB-40 aircraft, serial numbers, and names is included in Table I, along with a cutaway drawing in Figure I of the appendix.

The YB-40 was a specific modification of the B-17F and required my father to undergo additional instruction to become proficient and knowledgeable regarding the airplane before he could begin combat operations. The pilots selected to fly the secret YB-40 volunteered for that special duty, according to a letter written by Doc Furniss, the flight surgeon for the 327th BS.

Based on British losses in early bomber operations in the European Theater, due to the lack of escort fighters with enough range to go into enemy territory, the United States began the development of a secret armed escort bomber on June 25, 1941. The task was to convert a B-17F aircraft into a heavily armed escort aircraft, the YB-40, by providing additional guns and turrets, increased armor plating, and a maximum amount of ammunition. However, the YB-40 would not have the capability to carry bombs. It was developed purely to escort the B-17s all the way to the target and safely back to the home base.

The management of the 8th Air Force did not think, at the time, that fighter escorts would be necessary because the B-17s would be at a high enough altitude where flak would be ineffective. Therefore, they did not produce an effective fighter escort at the beginning of the war, and the powerful P-51 Mustang fighter had just started production. The P-51 was not available to accompany the B-17s deep into enemy territory until much later in the war.

The YB-40 had an extra gun turret on the top of the airplane above the radio room. In addition, the bombardier's equipment was replaced by twin machine guns located in a "chin" location directly below the bombardier's position in the forward nose. The chin turret on the YB-40 (on the experimental XB-40 aircraft) is shown in Figure 2-3.

Figure 2-3: Experimental XB-40 showing chin turret that was eventually used on YB-40 (USAF photo)

Instead of a single machine gun in each of the two waist gunner positions, twin machine guns were mounted. A picture of the extra turret on the top and the twin machine guns in one of the waist gunner positions is shown in Figure 2-4. As many as eighteen guns were operable on the YB-40, compared to thirteen on the B-17, with the bomb bay converted to carry ammunition. Rare schematics of the YB-40 are given in the appendix, Figure II, for reference.

Figure 2-4: YB-40 showing the extra turret on top and extra guns (USAF photo)

I had the honor of meeting navigator Robert H. "Bob" Doolan in Cincinnati shortly after his 105th birthday in 2022. Bob was in the 327th BS and flew seven combat missions as navigator in a YB-40. As there were only thirteen planes in the unit, he remembered my father. Bob told me that the top gun turrets on the airplane were ineffective because the arc through which they could rotate was very small—about thirty degrees. He also said that most of the YB-40 crews took out the armor plating in the YB-40s to lighten the load.

Upon learning that he would be one of the few pilots to fly the YB-40, my father was very excited and eager to fly the secret airplane. He and the other newly appointed pilots and crews of the YB-40s were off to the remote Biggs Army Air Field near San Antonio, Texas, in March 1943. After training, the crews began the long flight from El Paso to Orlando, my father's final cross-country journey before taking his YB-40 overseas from Florida.

The tests concluded that the YB-40 would be suitable for an escort aircraft for the B-17F. Although heavier, the YB-40 could keep up with the B-17s, even after they dropped their bombs and headed home. However, to keep up, manifold boost pressure had to be increased, and that tended to wear motors out more frequently compared to other aircraft. With that in mind, it was recommended that the YB-40 be used as the lead aircraft until just prior to the bomb run when it would trade positions with a B-17.

During their time in Florida, the YB-40 crews honed their navigation, flying, and formation flying skills. The YB-40s used the P-63 aircraft for "target practice," shooting dummy rounds at the P-63s. It is unusual to find a picture of a YB-40 in flight due to its classified nature, but Figure 2-5 shows a YB-40 on a training mission with the P-63s.

Figure 2-5: YB-40s on practice mission with P-63 aircraft that were used as "targets" (photo courtesy of 96th BG, Bombardier Fred Huston Collection)

Bob Doolan told me that because the YB-40 was a secret aircraft on a secret mission, it could not be kept at the airport at Orlando due to the visibility of the airfield. Therefore, the YB-40s were kept at nearby Montbrook Army Air Field in Williston, and the crewmembers stayed in tents on the base and cooked their own food.

Special Orders 118, dated April 28, 1943, from the USAAF School of Applied Tactics called for the dispatch of the thirteen YB-40s to England. The dispatch listed the serial numbers of the thirteen aircraft and the crewmembers who would ferry them overseas. My father's name was on that manifest, as was navigator Robert H. Doolan's.

Knowing that my father was about to embark on his flight overseas to fight in the war, my mother made one last trip to Florida to see him off. She flew on National Airlines to Jacksonville, Florida, (for $29.65!) and then took a bus to Williston, Florida, on April 26, 1943, as evidenced by the tickets in her scrapbook. They exchanged their tearful goodbyes on April 28, as he began the ferry flight of the YB-40 across the ocean. Mom's scrapbook also includes the train tickets from Orlando back to Dallas and the inscription, "Enroute home traveling companions: Helene Joy Liebman and a few tears. 4-28-43."

As they said their goodbyes, my mother and father did not know whether they would see each other again. At that time in the air war, loss of aircraft and aircrews was high because the Luftwaffe (German Air Force) fighters exacted a heavy toll on the vulnerable and unescorted bombers.

I am sure that my mother told my father the same thing she told me when I was selected to be an astronaut: "Put God first, then your country, and go for it." Then, the eighteen-year-old wept.

Chapter Three
Journey

The 92nd BG, nicknamed "Fame's Favoured Few," arrived in England in August 1942 and was initially based in Bovingdon. At the beginning of 1943, the 92nd BG moved to Alconbury and, by that time, was sending personnel and airplanes to North Africa to support the war effort there and to liberate the Mediterranean.

The 327th BS at Alconbury was utilized for training, passenger transport, and patrols given the limited manpower and planes that the 92nd BG initially received. Deployments were still being made to Africa, and trips to Gibraltar and Algiers were noted in the squadron reports. The 327th BS was anxious to get more deliveries of airplanes and personnel so that they could become operational and go to combat.

Alconbury in my father's day was a quaint village about thirty miles from Cambridge. Along with the 92nd BG, the 95th BG flew B-17s out of Alconbury, and the 93rd BG flew B-24 Liberators. At the time the 92nd BG arrived in Alconbury, there were only a few thousand troops in England, so there was a lot of excitement on the part of the Americans as well as the locals! A blow to the morale of the 92nd BG was the movement of several new B-17Fs and crews to the 97th BG to support their efforts in North Africa. In addition, the 92nd provided the necessary transportation of VIPs between England and North Africa. The addition of B-17s from the 8th BC added strength and air power that was needed to keep the Suez Canal and strategic Mediterranean ports under British, and therefore Allied, control.

Although there was much excitement in Alconbury about the potential to return to operational status, the base was not in a suitable condition for the number of aircraft and crews that would be coming. Construction of hangars had just started, roads were in the process of being marked, and there were not enough hard stands (concrete pads) on which to mount the aircraft, thereby requiring turf parking for the airplanes.

Most of the bases were temporary in nature and were carved out of farmland. When I visited the Alconbury air base in 2022, it was among gently rolling hills and charming English towns. By then, there was little remaining of the bases. It was hard to imagine the vibrant wartime bases when looking at the ghosts of their former selves. After the war, most of the bases returned to being peaceful and beautiful farms, but some were made into racetracks, government facilities, or prisons.

Living quarters at the base were not complete either and were, therefore, dispersed. Flying officers lived at a large estate (called Upton House) three miles away. They had to be transported by truck to and from the regular mess for each meal.

Over time, the hope to return to combat operations dimmed due to the slow arrival of airplanes and replacement crews. The higher priority was the war in Northern Africa, and other new aircraft that the 92nd BG received were diverted to the war in the Pacific.

Boredom in the 92nd BG ensued, even though additional courses in chemical warfare and camouflage were offered and plenty of work was needed to get the base ready. Basic training was mandatory with reveille at six every morning. The airmen had to march in formation to instill discipline. Other pastimes included competitive sports and dances, with baseball and volleyball being the most popular of the inter-squadron sports.

My father's journey to Alconbury in a YB-40 started on April 28, 1943, with a six-hour flight from Orlando to Mitchel Field on Long Island, New York. He chronicled his trip in a letter to his parents:

I had a rather interesting trip up here, stopping overnight at New York City. I went into town but didn't see much and didn't stay long. The city was dimmed out, so I couldn't see much. Flying up here, we flew over a lot of large cities—Richmond, Philadelphia, Boston, New York, Washington, D.C., etc. I saw the Capitol and flew almost right over it!

I bought some socks and underwear yesterday and a pair of G.I. brogan shoes! I also got a pair of gloves and a briefcase. (I'm really an operator of the large caliber now! I even have a .45 automatic and a hunting knife! I feel like a thug!)

After a second leg of the journey to the Presque Isle Army Air Field in northern Maine on April 29, 1943, he elaborated on the three-and-a-half-hour flight.

I arrived here at Presque Isle a few days ago, April 29, 1943, and will soon be on my way to England. Everything is in order for the hop, so I imagine we'll be on our way in a day or so.

My father called my mother when he was in Maine, and that was a difficult conversation because it was the last time they would be able to talk by phone before Daddy Ben went to Europe. As it turned out, they wouldn't have the chance to talk again for over two years.

From listening to other oral history accounts from B-17 pilots who ferried their "ships" to England, their feeling was one of wonderment and anticipation. Most of these young men had never been to another country, and they speculated what it would be like to fight a war. My father was no different, and I know that he was ready to use all of his training as a pilot to fight for our country and beat the Axis powers.

On May 2, 1943, Ben was busy getting ready to pilot his YB-40 "across the pond," albeit in frigid and uncertain weather conditions. His journal entry reflected his feelings:

I've been busy all day, getting the ship and everything ready to go. I test flew the ship this afternoon, and everything checked out fine. We're all ready to go now.

The leg on May 3 went from Presque Isle, Maine, to a Canadian-built base at Goose Bay, Labrador, and took three and a quarter hours of flight time. As my father wrote:

After a little radio trouble, we finally took off from Presque Isle, shortly after noon. Such barren country I've never before seen! It is mostly snow covered yet with no towns, roads, or such. We had no trouble and arrived here okay. It is cold up here, but not enough to offer discomfort.

Snow covers everything, reminding me of Spokane last December. It is interesting to note the icicles hanging from the eaves of the building—instead of hanging straight down, they curve due to the force of the wind.

As soon as we get a break in the weather, we'll be moving on, and it shouldn't be long.

The next part of the flight to Bluie West One base in southern Greenland was notoriously treacherous because it required the pilot to land on a glacier after flying through a narrow pass in the mountains. My father told of his journey, dated May 4, 1943:

We left Goose Bay about noon and had good weather most of the way. I was on instruments the last two hours though. This field in Greenland is tricky to get into, but we made it okay. The country around is barren and snow covered. To get to the field, we flew up a fjord at eight hundred feet, under the overcast, and between the mountains, fifty miles.

I saw the northern lights last night at Goose Bay, and it was a strange and wonderful sight.

The flight from Goose Bay to Bluie West One took over five hours. The weather in southern Greenland was usually bad and unpredictable, so I am glad that my father arrived there safely and that the weather was not "socked in" until after he had landed.

The next stop on the journey was to Iceland. According to my father's flight records, he flew from Greenland to Iceland, and the trip took four and a half hours. On May 6, 1943, he wrote:

I'm in Iceland now, and such a desolate place I've never before seen! It's very cold here, though there is no ice or snow on the island we are on.

We had a rather nice trip over but were on instruments most of the way, flying 1,500 feet above the water. The ceiling was about 4,000 [feet] *here though, so we had no trouble getting in. General Andrews was killed near here a day or so in zero weather* [zero visibility, zero cloud altitude].

Unfortunately, on May 3, 1943, the B-24 "Hot Stuff" carrying Lieutenant General Frank Andrews, the commander of all U.S. forces in the ETO, crashed into the side of a mountain while attempting to land at the Royal AFB in Iceland. Andrews and thirteen others died in the crash; only the tail gunner survived.

The final leg of my father's journey, occurring on May 7, 1943, took seven and a half hours, during which he traveled from Iceland to Scotland. Ben's journal documented that long flight:

We're in England at last after the longest leg of our trip. We flew instruments most of the way to Stornoway, Scotland, but were under the ceiling to Prestwick, where we landed. We ate there and then flew to our base here [Alconbury]. *It's rather nice here. The boys here were quite excited over our ships and are very nice to us.*

Indeed, the squadron was very happy to have its own planes arrive from the U.S. The entry in the 327th BS daily report on May 8, 1943, stated, "The long awaited YB-40 planes begin to arrive from the States, led by Major Robert Keck. These officers and men returned to the States early in the year to bring these aircraft to this [European] theatre. As they are still in the experimental stage, everyone is quite anxious to see the results. Each member of the squadron is elated over the fact that our squadron has been chosen to have these ships. From all reports, they will require lots of work from the engineering section and the flying crews will need quite a bit of familiarization. Everyone on the field has been warned how necessary it is to keep the facts about these planes a closely guarded secret."

There were thirteen YB-40s that left Florida on their way to England, but one YB-40 did not make it. On the last leg from Iceland to Scotland, it was forced to land in a peat bog on a Scottish island after running out of fuel. Although it was removed and repaired, it never flew in combat.

I am very proud that my father made it to England and that he was selected to be one of the few YB-40 pilots. Only twelve of these airplanes completed the trip overseas, and his was one of them. As a chosen "test pilot," he found his training and subsequent flight overseas to be challenging at best.

By the time my father arrived at the base in Alconbury, East Anglia, England, the facilities had improved to get ready for the new crews and airplanes; however, there was still work to do. My father's journal entry for May 8, 1943, the day after he arrived in England, stated:

We're living in a rather crude barracks, but we have the promise of much better quarters soon. I have my sleeping bag and some British blankets, so I keep warm at night. The food is very good.

It rains most of the time here and is still rather cold, especially at night. Everything is blacked out at night here [to prevent being seen for an attack].

I haven't talked with any natives yet, but they say it is amusing. I hope I can refrain from laughing at anybody, for I don't want to hurt anybody's feelings.

On May 9, my father had this to add:

Hugh and I went to town tonight, and what a quaint little town it is. The streets are very narrow, as are the sidewalks, and the buildings are all typically English, just like you see in the movies. We went to a movie, by the way, but they call them "cinemas" here. A lot of expressions are different like that. A drug store is a "chemist shop," and a hardware store is an "ironmonger." The people all seem friendly but in a distant sort of a way. You see very few cars here now, and the ones you do see are tiny ones—much like the Austin car in the U.S. Nearly all of the women are in some sort of a uniform.

Related to my father's discovery of the "proper" English language (versus American language), the *Instructions for American Servicemen in Britain 1942* book stated: "You will have to ask for sock suspenders to get garters and for braces instead of suspenders—if you need any. If you are standing in line to buy (book) a railroad ticket or a seat at the movies (cinema) you will be queuing (pronounced 'cueing') up before the booking office. If you want a beer quickly, you had better ask for the nearest pub." I think one of the most interesting admonitions was: "There's been a war on since 1939. The houses haven't been painted because factories are not making paint—they're making planes…The British people are anxious for you to know that, in normal times, Britain looks much prettier, cleaner, neater."

Until the base was operational, there was little that the newly arrived airmen could do for entertainment. Basically, they could cycle around town, the countryside, and the base, and there were trucks available for transportation. The USAAF personnel were required to always wear their uniforms, even during their time off and when they were not on the base.

Ben bought his own bicycle and wrote to his parents:

Yes, I'm getting a lot of good out of my bicycle! I haven't been "pleasure riding" for quite some time (it's too much work!) but I get plenty of riding. Nearly everybody over here has a bike due to the transportation problem. I'm getting my money's worth out of it, anyway, and I could hardly do without it!

A picture of some of the airmen at Alconbury with their bicycles and a few B-17s is shown in Figure 3-1.

Figure 3-1: Allied airmen at Alconbury ride bicycles on the field near the B-17 aircraft (courtesy of IWM Duxford Roger Freeman Collection)

The airmen were given a pass every two weeks, and Daddy Ben liked to go to London on leave, presumably traveling by train and staying in military accommodations. In addition to sightseeing, he went shopping. He bought more clothing in London, as it apparently took a long time to get cleaning done at the base. He also bought a better raincoat because he said that his "doesn't shed good enough"! I'm sure that weather in May was cloudy and rainy a good bit of the time, and, of course, that interfered with the ability to fly.

His journal entries further chronicled these days until combat sorties could begin:

May 10, 1943

Today has been "Gasless Monday" at our airdrome, and only the vitally necessary trucks have been running. There are usually trucks used as buses to carry us around, as our airdrome is scattered around. Since our drinking water is brought to our site in Lister Bags by truck, we have no drinking water tonight!

May 11, 1943

Bill [Stewart] *arrived today, along with most of the others. Casey cracked up in Scotland after being unable to land at Iceland. He made a belly landing in a peat bog with only five minutes of gas left. Tucker* is ten hours overdue somewhere up the line, so I suppose he has gone down somewhere, and I hope safe. None of Casey's crew were injured.*

**Tucker arrived okay a few days later.*

Bill Stewart was a good friend of Daddy Ben's, and they roomed together during their flight training. In a letter to his parents dated July 7, 1943, Ben described their friendship:

Bill Stewart and I are still running around together, and he's a real friend. We've been together ever since those first discouraging days at San Antonio, and I hope that we'll be discharged together! We go to the shows together, eat together, and during our spare time play checkers or solitaire together. His bed is next to mine, and we share a chest of drawers.

It's great to have a friend such as he is. He's a good, clean country boy with no bad habits to influence me with! We get along swell and have never had words during our year and a half of living together.

Upon the arrival of the new squadrons in England, there was a lot of indoctrination on the operational procedures used in the British Isles. Navigators (and pilots) had to become familiar with the British "G" or "Gee" radio compass equipment, the British "Splasher" system (beacons) for navigation fixes, and the new type of maps with which they would have to work.

In addition, bombardiers required new training on the targets they would have in Europe and became intimately familiar with landmarks and the countryside. Proficiency training was spent in the bombsight trainer, which simulated an actual bombing run. Radio operators attended code and signal classes, and gunners practiced in a simulator that had a dummy gun and a projected film of attacking fighters. The pilots also attended some of these classes and practiced formation flying as well as proficiency flying. Various emergency procedures, such as bailing out, air-sea rescue, evading capture, use of oxygen, and battle tactics, were practiced and re-taught as well.

On May 14, 1943, the 327th BS journal entry acknowledged, "The long-awaited day has finally arrived—we are once again an operational group. The changes in the personnel are very apparent since this has been announced. One can see it in the way the men walk to work. Everyone is hurrying. There is far less 'griping.' Maintenance crews have begun to work all kinds of hours—from early AM to late PM—and this is true of the other sections."

And so it began—the missions of the 327th BS of the 92nd BG from Alconbury, England. The 92nd BG would be denoted on the tail of the B-17s or YB-40s with the letter "B" in a triangle. The 327th BS designation on the fuselage and tail was "UX." Officially and proudly designated, the crews would finally experience war, and the eager ground crews assigned to each plane would effectively support the effort.

Chapter Four
Combat

There was no way that the men who had been sent to England to fly in combat or maintain the planes on the ground could anticipate the harsh conditions they would face. Through months of training and conditioning, there was nothing like the adrenaline rush of fighting and facing death with every flight.

The B-17 and YB-40 aircraft were unpressurized yet flew at altitudes of 25,000 feet or above. While the aircrews were on oxygen above ten thousand feet, the temperatures at those high altitudes were destructive to the human body in ways that aerospace medicine in those days did not understand. To combat the freezing temperatures that registered negative forty degrees or even negative sixty degrees Fahrenheit, some of the crew who were most exposed (side gunners or waist gunners) wore blue electrically heated one-piece undergarments called "blue bunny" suits. They were wired and attached to electrical outlets in the airplane. However, they could short out and also cause burns, and when they did not work, many crewmen got severe frostbite. Fortunately, the engines generated some heat, so those in or near the cockpit (pilot, copilot, flight engineer, navigator, and bombardier) could wear a leather flight jacket, uniform shirt and tie, khaki pants, fleece-lined boots, and wool gloves.

As much of the flying was over the water, each crewmember wore a life preserver and a parachute over that. A few locations in the plane for some of the crewmembers were too tight for them to wear the parachute, so it was placed nearby and hooked onto his harness in case of an emergency. In the speed and panic to abandon a plane quickly, often the parachutes did not get hooked onto harnesses correctly, saving lives but causing injuries. Finally, various headgear, microphones (including a throat microphone), radios, and headsets completed the flight suit. Metal flak helmets and vests were also added for extra protection near the target.

Not much has changed in what flight gear is needed for hazardous flying in airplanes or in space. My flight suit in the T-38 included everything that my father wore, except our cockpit was heated to the point that we did not need to have fleece linings in our flight jackets or boots! For my spaceflight missions, a much more substantial flight suit was needed in case we had to bail out of the Space Shuttle when a pressure suit was needed. In both cases, we called our life preservers "Mae Wests," so that name has stuck ever since WWII.

According to daily reports of the 92nd BG, they took off on May 14, 1943, with a mission to attack the major U-boat bunker and ship building facility at Kiel, Germany. German attacks came from Messerschmidt (Me) 109s and Focke-Wulf (FW) 190s, singly and in pairs. It was the first mission for the 92nd after becoming operational again as well as the first mission since my father had arrived in East Anglia.

I can imagine that it set a very somber tone at the base at Alconbury. Many of the new aircrews and ground crews had never seen formations fly off to combat and had never experienced the sense of loss and despair of not having an aircrew return home. The crew from the four airplanes from the 92nd BG on this first mission had quite a few harrowing stories to tell, and those opened the eyes of those who had not yet flown in combat.

On May 14, my father wrote:

They have started raiding from here, but we [the 327th] *won't be ready for missions for two weeks or so yet. It's a wonderful sight to see the sky filled with our bombers—B-17s. When the time comes for the raiders to return, we all gather on top of the control tower and count them as they peel off for landing. Then, we note the battle scars as they taxi by and offer a silent prayer for those who did not return. Soon, I will be going along, and I pray that after each mission, I will be able to "pass in review" for the control tower observers.*

When missions were flown, the ground rumbled all day, with airplanes taking off every twelve seconds and spiraling in vertical climbs into their formations. With the RAF airplanes flying at night, and the Allied airplanes flying during the day, the noise was constant throughout the formerly bucolic farmland.

After a day of ground school on May 17, my father and some of his friends were able to relieve some stress and take bicycle rides through the English countryside on May 18. I can't help but think that the advice in the book *Instructions for American Servicemen 1942* was on their minds. For example, "The British ...great place of recreation is the 'pub.' A pub, or public house, is what we could call a bar or tavern. The usual drink is beer, which is not an imitation of German beer as our beer is but ale...The British are beer drinkers—and can hold it. The beer is now below peace-time strength but can still make a man's tongue wag at both ends. You will be welcome in the British pubs as long as you remember one thing. The pub is 'the poor man's club,' the neighborhood or village gathering place, where the men have come to see their friends, not strangers."

Opting to skip the pubs that day, they enjoyed the beauty of the English countryside. As my father recounted:

Bill, Hugh, and I went cycling today through the country. We passed through several small villages, and they all look practically alike. All of the buildings are of brick, and many of them have thatched roofs. There are very few yards, and the ones you see are usually just small flowerbeds. The country is like a large green carpet, splotched with tree groves and hedge rows. We enjoyed the ride and plan on going again.

On May 19th, 1943, the 327th BS had its first combat mission since becoming operational in May. One of its YB-40 pilots, Captain Roland Sargent, acted as a copilot in a B-17 for the 325th BS on another mission to Kiel to attack the turbine engine building installations. The weather was perfect, and all of the targets were bombed. It is interesting to note that, particularly around Kiel, the Germans used smoke pots to create heavy smoke so that the targets could not be seen.

One of the aircrew members on his first trip remembered, "It was just about what I had expected from the reports that crews who had been over before had given me. What I wanted most when we left the target area was a big bottle of Budweiser beer. Man, was I dry."

The 327th BS was the only outfit in the entire USAAF to have YB-40s, and on May 29, 1943, their first mission was recorded. "Our planes (the new YB-40s) were dispatched on their first combat mission today. They were ordered to accompany the B-17s of the group to St. Nazaire [France]. They returned without incident as the expected fighter opposition was not encountered. After their first trip on these ships, the crews are confident that the ships will be able to do all that is expected of them, and they are anxious to have another chance at 'Jerry' [slang for the Germans]."

Seven YB-40s were deployed on that mission to escort the fifteen Flying Fortresses from the 325th and 327th squadrons, and they had no difficulty maintaining formation with the B-17s. From all aircrew accounts, it was a successful mission with no casualties and very little damage to the aircraft. The damage that did occur came from flak, not from enemy aircraft.

The airmen who had never flown in combat probably gathered around in the base bar and listened to the stories of those who had just come back from their missions. They learned that there were some fighter escorts available, the Thunderbolt P-47 aircraft, known as the "Jug," that had recently been fitted with long-range drop fuel tanks. However, the P-47s, because of their limited range, could only escort the bombers to the coastline of enemy territory before they had to return to England. The German fighters knew to wait for the escorts to leave, and then they attacked our bombers.

The first combat mission for the YB-40s showed their deficiencies. There were armament flaws, modifications of the waist and tail gun feeds to be addressed, and more ammunition supplies needed. Therefore, they were grounded from May 29 until June 15 to correct these issues. A complete list of the nine missions on which the YB-40s were deployed is summarized in Table II in the appendix.

My father was disappointed that he was not on that first combat mission of the YB-40s. On that same day, he was getting his YB-40 prepared for combat. The name of that ship was "Dollie Madison" (serial number 42-5744). In one of his letters to home, he stated that he always felt like the plane was "Dolly Jo." However, to avoid directly attributing the name to my mother, I guess he found a very close alternative! Generally, the pilot was able to name the aircraft, but I'm sure that he would only pick a name that everyone liked.

On May 29th, Daddy Ben's war diary was comprehensive as it discussed day-to-day life on the base as well as the terror of missions:

I've been busy all afternoon getting our ship loaded with ammunition and firing in the guns. I expect to be in action any day now. I was scheduled for a mission tomorrow, but it has been scrubbed.

As combat time neared, pilots sharpened their skills, as did the navigators, bombardiers, and gunners. There was practice on the difficult task of spiraling into formations and how to join other squadrons. Assembling hundreds of airplanes and integrating with other squadrons took extreme accuracy and planning. Many men died just getting into formation, never having seen a battle.

On June 5, Daddy Ben flew the YB-40 for almost five hours on a training mission to get the plane, his crew, and himself ready for combat. The men always fired their guns after crossing the British coastline on the way to Europe to make sure they worked. As the day drew near for Dollie Madison's first journey into combat, a few more checks were necessary, as related in my father's journal on June 16:

We fired our guns on the ground today, and they are in fairly good shape. Just a few minor adjustments, and she'll be ready for combat, and I imagine we'll be taking her to the ball game soon.

My father had one thing on his mind: to be ready to fly into combat. He was ready, his crew was ready, and his airplane was ready. It was time to go to war.

Ben's First Combat Mission
June 22, 1943

The young men who had gone to war were brimming with bravery and excitement. They wanted so badly to fight for our country and for the Allied forces to win the conflict against the Axis countries. These airmen in England had been getting ready for months, practicing and training to fly in combat or to prepare the bombs, armament, airplanes, facilities, and whatever was needed to fight the enemy.

On June 22, 1943, my twenty-four-year-old father arrived at the long-awaited day when he was going to fight in combat. As was typical for those missions, the assigned crews were awakened at a very early hour, ate breakfast, and then went to the briefing room to learn about the impending mission. Even before that, the ground crews and gunners were preparing the aircraft as well as loading ammunition and bombs. The planes on their hard stands were ready to take flight in their role as Flying Fortresses.

Figure 4-1: Briefing room at Alconbury AAF base (photo courtesy of Roger Freeman Collection, IWM Duxford)

In the briefing room (Figure 4-1), there was a large map on the wall behind a curtain. After the men filed in, with the top officers coming in last, the briefing began. The curtain was drawn back, and behind it, the map revealed the target(s) and the paths to get there marked with a red ribbon. The pilots and aircrews were also briefed on the other bomb groups participating as well as formations, time hacks, and codes (for initial points, rally points, altitudes, beacons, etc.) Usually, there would be subsequent meetings for the pilots only, navigators only, and bombardiers only. According to journalist Andy Rooney, they were all scared. If they didn't admit to being scared, they were lying.

I often wonder what was going through my father's mind as he awoke early in the morning and sat through the briefing for his first combat mission. I'm sure he was excited, perhaps apprehensive—even scared. I imagine he had some of the same emotions that I experienced the morning of my first spaceflight.

Ours was the first class of astronauts selected after the Challenger accident. After eating breakfast that morning and donning my launch and entry suit (LES), I was ready to go. After five years of training and preparation, I wanted to go to space and do my job. I admittedly had the jitters, mostly because I did not want to make a mistake and do anything to jeopardize the mission—that was my main concern. I also wasn't sure if I would get space motion sickness that would sideline me for a couple of days. There was no way to prepare for this or know how my body would react. My solution was to have a chocolate breakfast shake, which I figured would be less disgusting if I saw it again.

As my father rode out to his airplane with his crew, and as I rode out to the Shuttle launch pad in the Airstream Astrovan with my own crew, I suspect that we had some of the same feelings. We were serving our country out of patriotism and a sense of duty. The fact that we were participating in highly risky missions was not forefront in our minds. We were doing what we wanted to do and what we were trained to do. We had done everything that we could to prepare for our respective missions, and we relied on the capabilities and professionalism of our ground crews and support team. We had long ago accepted that death was a possible outcome, but we did not dwell on it or let it deter us from doing what we felt we were called to do.

As June 1943 was early in the air war in the European Theater, daily bombing missions concentrated on destroying manufacturing and industrial plants, supply chain plants for building armament and aircraft, and submarine pens. On this day, the critical target was the IG Farben synthetic rubber manufacturing facility in Hüls near Recklinghausen, Germany, just north of the Ruhr area. This plant was the largest of its kind in Germany, and this mission was, to date, the closest approach of the USAAF to Germany's heavily defended Ruhr industrial region. The secondary targets were the Ford and General Motors plants at Antwerp, Belgium, and there were diversion missions to shipyards at Rotterdam and to the North Sea. The diversion and secondary target missions were designed to draw enemy aircraft and resources away from the primary target.

The comprehensive 8th BC reports of each of its missions included a narrative summary, enemy aircraft and flak summary, bombs deployed (and their accuracy), Allied losses in terms of aircraft and personnel, photographs, and problems, among other things. I was able to obtain these records for all my father's missions thanks to the fine folks at the Air Force Historical Research Agency at Maxwell AFB. The report for each mission was literally hundreds of pages long. It is hard to imagine how they were generated almost daily with typewriters, carbon paper, and handmade charts and drawings. Also made available to me were the daily reports of the 92nd BG as well as the daily operations reports of the 327th BS. A summary of my father's combat missions, along with a map of the target locations, is given in Table III and Figure III in the appendix.

Early in the air war, the USAAF was struggling with bombing accuracy and trying to devise combat boxes for the bomber formations so that they would not be as vulnerable to the enemy. The enemy aircraft would typically attack the Allied aircraft from the front or the rear because the nose area and the tail were most vulnerable, as was the lead aircraft for the mission. In fact, the latest tactic at the date of this mission was for the Luftwaffe to shoot down or cripple the leader of the formation. The enemy aircraft would fly head on and then do a half-roll just after attacking.

According to Ben's wartime journal:

I went on my first raid today as copilot for Captain Roland Sargent. The target was a synthetic rubber plant at Hüls in the Ruhr valley. They woke us up at 3:30 a.m., and we went to the mess hall and ate breakfast.

Then, we went to the briefing room at 5:00 and were briefed on the mission, which was a "maximum effort." We then got our flying equipment from our lockers and went to our ships. We were flying my old ship [Dollie Madison]*, which is still "Dolly Jo" to me.*

One thing my father's first combat mission and my first Space Shuttle mission definitely had in common was that Dolly Jo was praying for us as we entered a treacherous domain. It is much harder for spouses, parents, friends, and family to watch someone they love face danger than it is for those in the cockpit.

On this early morning in June 1943, Ben approached his "ship," as he called it; threw his flight gear into the plane; and swung into the door under the front of the aircraft by grabbing the opening and hurling his legs and body up into the nose. Almost fifty years later, after disembarking the Astrovan, I awkwardly waddled to the launch pad elevator in my partial pressure suit as other support personnel carried the stuff that I would need for comfort and survival. The Space Shuttle creaked and groaned like never before because it was seething with the liquid oxygen and liquid hydrogen that had been loaded into the vehicle the night before. It was alive! Ben's vehicle was alive too—that day, the amount of armament was different from anything with which he had ever practiced.

Decades apart, the copilot in "Dollie Madison" and the astronaut in "Endeavour" busied themselves with completing checklists, verifying configuration, strapping in, hooking up oxygen, checking communication, and working with their crews to make sure that everything was ready. We were both focused and excited, and that helped to ease any butterflies in our stomachs. I think we experienced a lot of the same thoughts, fears, and feelings of determination and readiness. These were different missions—I wasn't going to be shot at, and he was not sitting on top of a seven-million-pound controlled explosion. But neither of us fixated on the danger ahead. The "Dollie Madison" crew of ten and the Space Shuttle "Endeavour" crew of seven were ready to go.

At the signal of a white flare, the pilot and copilot started each of the YB-40's four engines, and my father was surely comforted by the familiarity of his ship as he mentally reviewed the details of the briefing. Each engine awkwardly started with a puff of smoke and drops of oil as Dollie Madison sprang to life.

My father was not allowed to write anything down from the briefing unless it was coded and unrecognizable in the event he was shot down or had to bail out of his aircraft. However, he later wrote in his journal:

We took off about 7:00 and flew to another field to join a group there. Then, we flew around England, joining other groups until the formation was made up. It was a large formation and beautiful to look at.

Individual BGs were assembled over their respective bases, and combat wings (CWs) were assembled over designated points. Once aloft, "lead ships" with colored flares would direct the bombers to predetermined points where they would organize themselves into their attack formations. These groups of large and loud aircraft proceeded over the North Sea, climbing to respective bombing altitudes before crossing the enemy coast. It took about an hour to gather all these formations together, and it must have been a spectacular sight to see all of those aircraft in flight.

For this mission, the first CW formation consisted of B-17s and YB-40s from the 92nd BG and was designated the lead at a cruising and bombing altitude of 26,000 feet. The lead airplane of the high formation had the crew who was responsible for the navigation of the entire group of 160 aircraft and for calling "bombs away" over the primary target. Of course, the other bombers were expected to follow that lead. A photograph of a formation of B-17s from the 92nd BG dropping their bombs is shown in Figure 4-2.

The Flying Fortresses had to fly flat and level for two to three miles before bombing their target to get the Norden bombsight to work correctly. The Norden bombsight was one of the most technologically advanced devices of its time and was used to precisely determine the location of the bomb's impact point. It could calculate the aircraft's ground speed and direction and perform as an autopilot to accurately fly the plane to its target.

As the YB-40s were escort aircraft for the B-17s, they were in front and alongside the lead group. As the formation approached the target, the YB-40 in the front would switch places with the lead group's primary

B-17. Putting this airplane at the highest position was thought to make it more resistant to flak and enemy fighters who attacked head on. Also, the dynamics of flying at a higher altitude gave them a higher airspeed, which reduced "jamming up" over the target. In this mission report, the formation was called "stepped down."

My father continued his description of that harrowing day:

Nothing unusual happened until we had crossed the coast, heading for Germany, but then we got some light flak and, shortly after, fighters. Somehow, the fighters didn't bother me much, though I could see them shooting at us. I guess I was too busy thinking of flying. The fighters kept attacking, and as we passed over the target, I saw two forts go down.

Figure 4-2: Bombing run of B-17s from 92nd BG, September 1944 (USAAF Photo)

The Germans designed the 88mm anti-aircraft flak guns after World War I to be mobile artillery. Using sound detection and radar, the flak guns could detect oncoming aircraft and automatically adjust their azimuth and elevation (direction and angle from the horizon) so that the shells would explode at the targeted altitude of the aircraft. When exploded, the shrapnel from the twenty-pound projectile could pierce almost five inches of armor. Therefore, the flak guns were very effective and invoked a lot of damage on aircrew and aircraft. An example of a flak-damaged aircraft from my father's squadron is shown in Figure 4-3.

Figure 4-3: Severe flak damage of B-17 from 92nd BG 327th BS, August 1944 (USAAF Photo)

Several B-17 crewmembers remembered the puffs of black smoke and the shuddering of the aircraft hit by the flak. They commented that the flak was so heavy "you could walk on it." General Eaker required that airplanes remain in formation, even when flying through flak, so evasive action to avoid it was not possible.

It was reported that opposition was strong, with more than 125 enemy aircraft encountered for those aircraft going to the Hüls target. The tactics by the enemy varied from single to formation attacks of three or more aircraft. The formation attacks were all high and head-on between the eleven and one o'clock position. Single attacks covered all positions of the clock with the greatest number coming from the rear at six o'clock and from below.

After we turned off the target, one of our YBs, Bill Carey, went down slowly with #4 engine feathered and #1 smoking. We were in flak in the target area.

The YB-40 that was lost was piloted by First Lieutenant Andrew F. Bilek and copiloted by First Lieutenant William P. Carey, as well as eight other crewmembers. The aircraft "Wango Wango" (42-5735) was last seen going down under control near the Dutch frontier. It had been hit by flak. One engine was on fire, and it crash landed at Pont, sixteen miles northwest of Krefeld, Germany. It would be the only YB-40 lost in the war, as the YB-40 program wound down shortly thereafter. All ten crewmembers of Wango Wango bailed out and became POWs in the German camps.

At that point, the bombing was over, and the B-17 and YB-40 aircraft headed back to East Anglia. I don't know whether Ben felt a sigh of relief, as there was still danger ahead from flak and German fighters. It must have been awful to see one of his fellow YB-40s and its crew go down, especially considering there were only twelve YB-40s in the European Theater. After your friends and their airplane are lost, you consider your own vulnerability, but you must put it behind you, regroup, and remain vigilant.

Those feelings related to the loss of friends and beloved aircraft are similar to those experienced in the space program. After the tragic accident of the Space Shuttle Challenger in 1986, we all had to face the real dangers of going into space. I knew some of the people on board the Challenger, and I was very involved at NASA with returning the Space Shuttle to flight status.

On the day of the Challenger accident, Mom called me to comfort me and ask whether I still wanted to be an astronaut. Without hesitation, I responded, "Absolutely!" The accident and the loss of seven crewmembers did not deter me—I knew that flying in space had its dangers. Likewise, even after he saw the YB-40 spiral down in the war, it did not deter my father from flying again. He knew that flying in combat was perilous.

While returning to England from the target in Germany, the friendly fighter escorts picked up the formation at the Dutch coast and stayed with it until the English coast.

The [enemy] *fighters stayed with us, and just before reaching the coast, some English Spitfires met us right on schedule. About that time, a Focke-Wulf 190 scored two direct hits on us with 20mm cannons, knocking a hole in the leading edge of the left wing and knocking off the radio turret, wounding the gunner. I was flying at the time, and the ship lurched, started vibrating, and veered off to the left. The Spits came in to cover us as we lost formation. The ailerons were fluttering, and the left wing tip was waving, and we didn't know whether or not we'd make it. We kept going, though, after reducing speed and made the English coast okay.*

Another accounting of my father's first combat mission was sent in a letter to his family:

Well, I'm no longer a "virgin," for I received my baptism of fire yesterday. I went on my first raid, and it was quite a show… We had to drop out of formation, but there were some English Spitfires around to protect us luckily! The Spitfires escorted us all the way back to England, and they really looked good out there! We didn't know whether or not the wing was going to come off, so we sweated plenty on the way back; the English coast really looked good! We landed with a flat tire without mishap.

In the debriefing, Staff Sergeant Charles T. Foster of Harlan, Kentucky, right waist gunner of the B-17 "What Next," said while describing the excellent fighter cover: "I saw two Spits converge on an FW, and they really got him, and I don't mean maybe. He was trying to get to the ground away from the Spits, but he was very unsuccessful. They caught him when he was about three thousand feet up. It was the nicest example of crossfire I have ever seen."

Another aircraft had a feathered propellor and a smoking second engine when the Spitfires joined the B-17s. They protected the B-17 ship so well from the attacking Germans that the copilot, Second Lieutenant William R. Fellenbaum of San Bernardino, California, made this comment: "The FWs were queued up on us, ready to come in for the kill, when, all of a sudden, the Spits came out of nowhere. It sounded like they were playing 'The Star-Spangled Banner.' Then, the FWs scattered like flies with the Spits in hot pursuit. I really never expected to come back until this happened, and now I believe in the impossible." Of particular interest to the 8th BC was the inadequacy of fighter escort aircraft in the USAAF fleet. To supplement the USAAF aircraft, the RAF provided Spitfires as an escort to bomber aircraft formations returning to England. The 8th BC report exclaimed that the fighter escort coverage had improved with every mission.

In the journal to my mother, my father wrote:

We landed with a flat tire. We were both on the controls all the way across, as it was difficult to hold the ship level. I went with Sgt. Fogarty to the sick bay and helped the doctor a little. Another B-17 had made a belly landing ahead of us with five wounded by flak. Another squadron sent another ship to pick us up, and we got home about 5:00 p.m. Everybody was glad to see us, but not half as glad as I was to be back! The Lord answered my prayers and brought us safely home.

In its summary report, the 8th BC declared the maximum effort bombing mission successful over the primary target of Hüls, with the primary power station being hit as well as three large gas holders near the center of the plant. The secondary target at Antwerp was also successfully bombed. However, the hierarchy of the 8th BC was still hoping for more precise bombing and put plans in place to increase accuracy of bombing runs. It usually took several raids to the same target before the bombing was complete, so there were plenty of opportunities to improve.

There were some positive lessons learned from this mission about communication between aircraft in formations as well as fighter support. Although this mission was deemed successful, it was not without loss. Twenty of the heavy bombers were missing (sixteen over Hüls and four over Antwerp), and several were damaged severely.

As an update on Ben's injured crewmate, he entered this information in his journal on July 6, 1943:

I just saw T/Sgt Fogarty, our gunner who was wounded in the Hüls raid. He wasn't wounded as badly as we thought and is ready to fly again now.

My father was in the air war now, and he logged five and a quarter hours of combat flying time that day. He had a memorable first combat mission with the loss of friends and aircraft in enemy territory. Not only did he witness the complete loss of B-17s with their crews, but he also experienced one of his crew being injured. In addition to this emotional trauma, his aircraft was heavily damaged and limped back to England. Ben did not know whether or not he would make it back to a remote base. The flak and enemy aircraft attacks had been harrowing, and his adrenaline had surely been pumping while he tried to fly the airplane and stay in formation. His gunners had been freezing, exposed to the temperatures at the high altitudes, as he did his best to keep the "Jerries" at bay. It was a mission that my father would never forget, and one that would change his life forever.

This first combat mission on June 22, 1943, must have weighed heavily on his mind as he sketched a pencil self-portrait on June 24 after that challenging day (Figure 4-4). In the drawing, he looked tired and haggard, no longer a novice aviator but a more mature and war-torn bomber pilot who had just flown in combat. Although exhausted and war weary, his confidence was bolstered since he and his crippled plane and injured crew had survived. And Dolly Jo's prayers were answered.

Figure 4-4: Self-portrait of Ben Smotherman after his first combat mission, June 22, 1943

Ben's Second Combat Mission
June 29, 1943

In between training flights and combat missions, my father relaxed by drawing and painting, something that had brought him joy from a very early age. In a letter to his parents, he wrote:

I've been keeping in practice with my cartooning by drawing a strip, and I have the boys over here quite interested! They come around every day and say, "Can I see today's funnies?" Also, I'm decorating the walls of the pilot's room, though I haven't done much on it yet. So, I'm keeping in practice in case I ever get to be a cartoonist.

In another letter, he mentioned that he was painting life-sized girls on the wall of the 327th BS Pilot's Room. Then, on June 27, 1943, he stated that it was pretty tiresome but that he was almost finished!

By mid-1943, more buildings had been erected at the Alconbury field, including the hastily built Nissen huts, which were prefabricated steel structures for military use. Primarily used as barracks, they were made from a half-cylindrical skin of corrugated steel, and in the U.S., they were known as Quonset huts.

When I visited several WWII airfields in East Anglia, some of them had the original Nissen huts with paintings on the walls. I imagined what my father's paintings in his buildings had looked like, but unfortunately, none of those buildings at the Alconbury field are still standing.

Ben had to abort his next mission to St. Nazaire on June 28, as he recounted in his journal:

We started on another raid today in a B-17F but had to leave the formation and return home before leaving the English coast. The prop governor on #2 engine broke, causing the engine to turn up too fast. The raid turned out to be highly successful, and I wish we could have completed it. Jim Combest, a navigator whom I have flown with several times in the

States, got killed by flak. He is the first of our bunch that came over to be killed in action. Combest was in another squadron, having transferred.

After jettisoning his bombs over the English Channel, Ben Smotherman returned to Alconbury (Figure 4-5). He was one of the officers on the ground who faithfully and anxiously watched their colleagues return from France. Fliers and ground crew usually gathered at the block house to watch the skies above and count as each plane, many battered and with injured aboard, returned to the base. The approaching planes signaled with two red flares if wounded were aboard, and ambulances were summoned and stood by with medics to meet the planes.

Figure 4-5: Ben Smotherman's B-17F during his aborted combat mission to St. Nazaire, June 29, 1943 (USAF photo)

One of those red flares was from the B-17 carrying James Oliver Combest, a friend of my father. Combest was killed when he was hit in the abdomen by a large chunk of flak. There was no first aid possible, and he died within thirty seconds of being hit. He was from Kentucky and died at the age of twenty-six. He was awarded the Purple Heart posthumously.

On the aborted mission to Nazaire, Ben logged two hours of combat time in a B-17F, but I know he was disappointed that he could not complete the mission. That particular mission was very tough on my father, as his buddy Combest was KIA, and it reminded him of the nature of war and the price some had to pay.

The next day, my father was assigned to fly on June 29, 1943. The primary target was Villacoublay in northern France near Paris. The location housed a military airfield for the Luftwaffe in occupied France. Seized by the Germans in June 1940, Villacoublay was used as a Luftwaffe military airfield during the occupation. Notably, the airfield at Villacoublay has an interesting history, as it was used to show off the Wright brothers' biplanes in the early 1900s. Count Charles de Lambert owned two Wright biplanes and set up the Wright-Astra flying school there.

The 92nd BG sent fourteen aircraft, of which two were YB-40s. There was also a diversion operation involving forty aircraft that was successful in diverting the attention of the Germans.

Unfortunately, the weather in England was constantly an issue, with frequent rain and fog interfering with formations as well as takeoffs and landings. The weather had to be suitable to take off from airfields in East Anglia, and the bombardier had to be able to see the target in order to accurately drop bombs.

Weather and clouds at the targets were a chronic problem. At the target area near Villacoublay, there was a solid cloud layer with tops estimated to be at four thousand feet. As a result, the primary target was not attacked, and Ben's B-17 was one of those aircraft that returned to the base without dropping any bombs. However, as he flew into enemy territory, it was categorized as a combat mission. He recounted in his journal:

I completed my second raid today, but it wasn't very eventful. Our target was at Paris, but a solid overcast prevented our bombing, so we brought our bombs home. I was in a B-17 again. We had fairly stiff fighter attacks but only light flak. All of our ships returned safely. Our crew claimed two fighters down with one probable.

For this mission, eight squadrons of P-47s furnished aircraft to escort the bombers on their return to England from Villacoublay. Squadrons of RAF Spitfires also furnished withdrawal support for the secondary effort without incident, although twenty to thirty enemy aircraft were observed.

The 8th BC was very complimentary about the fighter escort support. In a follow-up report, it stated: "It should be emphasized that the fighter escort for our bombers, which is furnished by both 8th Air Force Fighter Command and RAF Fighter Command, is becoming more and more effective. The aim of fighter escort is to prevent attacks by enemy aircraft upon our bomber formations and not to shoot down enemy aircraft after such attacks have been made. The fighter coordination has improved to the point whereby this aim is being realized; enemy fighters are not attacking our bombers once friendly fighter support is picked up."

The June 29th mission demonstrated that further coordination by the bombardier and navigator was important. Navigators precisely pinpointed the start of the bombardier's bomb run, which enabled the bombardier complete freedom in his attempts to pick up the target in his sight. After the navigator successfully located the target, the bombardier took over control of the plane from thc pilot and bombed the target. Therefore, communication between the various members of the aircrew was essential.

This point was further emphasized by the commander of the Air Corps, Colonel Curtis E. LeMay, who stated: "This mission brings out more strongly than ever the importance of target identification and the need for continued emphasis on this subject in the ground training program. All bombardier-navigator teams must be thoroughly familiar with the target and the surrounding terrain. It is impossible in the short period of time allowed during [the] briefing to gain the degree of familiarity required. Hours of study must be made in ground school and the briefing used only for review." Clearly, some bombardiers and navigators had some homework to do after this mission!

Between missions and during stretches of bad weather, there was time for letter writing. Ben indicated more self-confidence regarding his missions. In a letter to his parents dated June 30, 1943, he shared:

I'm getting along just fine, folks, and feeling fine and dandy. My morale is high, and I don't have war nerves! This gang I live with is just like a bunch of college boys, and there's always some kind of foolishness going on! I keep busy most of the time, so the time passes fairly fast. I still like it okay here, though the "new" has worn off, and it doesn't seem so unusual here anymore. I've got to where I can understand the British, so I get along. I'm taking care of myself, and I'm glad to say that I'm in good health. I feel as good as I ever have. I have a very good appetite, so I eat good, and I get plenty of rest and exercise. So, try not to worry about me, for I'm doing fine.

Coincidentally, the 327th squadron party was also held on June 30 at the Red Cross Aero Club on the base. I don't know if my father felt like partying, but it might have been a welcome relief to have a little fun. He mentioned in this same letter that there was a party but that he didn't think that he would stay for it. However, he did say that he would go down for the "eats" because they had steak and ice cream! Hopefully, everyone had a good time despite the war around them.

Ben's Third Combat Mission
July 4. 1943

During the first few days of July 1943, the weather did not allow any combat missions. However, it gave the ground crews needed time to perform maintenance on the aircraft and gave the pilots a break from combat and allowed them time to do some local short flights, or "hops."

On July 2, Ben was scheduled on a mission to Nantes, France, as he wrote in his journal:

We started on a raid today but were called back as we started to taxi. It looked like an easy one, though long, and I wish we could have gone on.

That same day, my father received some pictures from my mother, which made him very happy. After being away from her for two months and being in the middle of combat operations, he was homesick and missed her a lot. He wrote that he didn't remember ever being so happy to receive pictures!

On July 4, when most people back in the United States were celebrating Independence Day, there were thousands of soldiers fighting the war on multiple fronts. My father celebrated Independence Day by going on another combat flight, this time to Nantes, France. He participated in firecrackers of a different kind—liberty bombs! In keeping with the USAAF policy of precision bombing of industrial targets, Nantes was targeted due to its Société Nationale de Constructions Aéronautiques de L'Ouest aircraft factory at Nantes Château Bougon. It was a heavily industrialized area subject to more bombing as the war wore on.

As an example of the increased expeditious pace of the missions, the bomb loading orders for the Nantes mission were received by telephone at 5:30 p.m. the day before. The attack order was received from the headquarters of the 1st Bombardment Wing close to midnight. After a very early wakeup for the flight crews, the briefing was held at 7 a.m., and the crews took off at 10 a.m. The five-hundred-pound bombs and armament had been loaded even earlier by the ground crews and gunners.

It was a very long mission, but there were several air bases in East Anglia that were available for refueling, medical support, and emergency landings. In addition, the sea rescue for downed aircraft was ready on high alert. Those kinds of plans were prepared for any mission, but when a mission was a lengthy one that required flying over water and going deep into France, those preparations were all the more necessary. Planning for emergencies was key. Some of the missions deep into enemy territory could last eight hours.

The target at Nantes was reached at about 12:45 p.m. at 22,000 feet. The B-17s of the 92nd BG flew in the low squadron position in the box formation. However, each of the three YB-40s was assigned as escort for different groups. One of the YB-40s was Peoria Prowler (42-5733), and one was Chicago (42-5741), but reports did not identify the third YB-40 on that mission.

Formation flying at such long distances was difficult. From my own experience, to maintain formation with our NASA T-38 jets, the pilot had to constantly adjust the throttles to maintain horizontal position as well as altitude. The T-38 had two engines, so the throttles were relatively easy to control together. Having to control four engines with four throttles would compound the difficulty! I am sure it was a two-person effort in the B-17 most of the time.

On that day, my father wrote:

I went on my third raid today, Nantes, France, and it was a long one—seven hours. We had no trouble until we hit the target, when about fifty fighters hit us. We had a running battle for about fifty miles and got several holes in the ship. Our tail gunner got slightly wounded in the leg but not serious. Campbell went down over the target.

The tail gunner on my father's crew, whom he mentioned as being wounded, was Staff Sergeant John C. Ford. My father also witnessed the B-17 that went down carrying pilot Lieutenant John J. Campbell and his crew. They were seen bailing out of their burning plane (42-29967) that crashed into the water. All ten crewmembers survived to be "guests of the Luftwaffe" in their prisons.

All reports indicated that bombing results at Nantes were excellent, with successful bombing at the aircraft factory and railway. The targets were repeatedly pounded. They received much damage due to accuracy of the bombing, slowing future assembly of aircraft and disrupting vital transportation links.

Enemy air opposition was moderate to strong, with an average of seventy to ninety enemy aircraft for each effort. My father was not exaggerating when he estimated that approximately fifty enemy aircraft attacked him near the target! Once more, FW-190s and Me 109s were the predominant enemy aircraft. Although most enemy attacks were from the front of the bomber, attacks on the rear also occurred and caused the injury of the tail gunner in my father's aircraft. After the attacks, the enemy fighters disengaged by executing a half roll, breaking down and under the bomber.

On a brighter note, the war diary of the 327th BS noted on July 9 that, "Our squadron day room has now been completely furnished for the use of the men. A large radio and Victrola combination set has been purchased out of the squadron fund and is the center of attraction in the day room." Having the ability to listen to music in their time off was a big morale booster given the stress and strain that the flight crews, ground crews, and senior officers were experiencing.

Ben's Fourth Combat Mission
July 14, 1943

Villacoublay was once again the target for Ben Smotherman's next combat mission on July 14. The mission had been attempted on July 13 but had been scrubbed due to weather.

That day, the 92nd BG sent ten B-17s and five YB-40s. Bombs were dropped on the aircraft repair and assembly shops, hangars, bomb storage area, workshops, aircraft, servicing tarmac, and main runway. The 8th BC reported that bombing was excellent, which validated the strategy to perform precision daylight bombing against industrial targets in the occupied territory.

It is not clear whether my father was flying a B-17 or a YB-40 that day, but he related:

I went on my fourth raid today to Paris. Our target was a Focke-Wulf repair factory. We had a fighter escort going in and coming out. The flak was pretty heavy, and there were quite a few fighters.

As for us, we lost #1 engine just before leaving the French coast but suffered no other damage and stayed in formation with #1 feathered.

When one of the four B-17 or YB-40 engines quit working, the blades of the propellor of the failed or shutdown engine could be rotated so that their outer section aligned with the airflow. This practice, called "feathering," created less drag and air resistance, thereby conserving power and fuel consumed by the remaining engines. It is amazing that Ben's aircraft was able to stay in formation, as the other three engines had to have increased power in order to maintain airspeed.

The strategy of the Allies' daylight precision bombing of airfields and aircraft manufacturing plants was successful. However, the German industrial machine was sufficiently geographically distributed for the bombing to not have a significant impact on the production of airplanes. As a result, Hitler and the Axis powers were able to bounce back quickly and to stay in the air war. In addition, slave labor, often done by women, was utilized to repair marshalling yards immediately, keeping transportation intact.

For the Villacoublay mission, all ships returned safely with only two slightly wounded men. However, there were eight B-17s lost from another BG as well as two aircraft from the Spitfire squadrons. The losses of B-17 crews, whether missing in action (MIA) or KIA, were fairly high by then. There are estimates from 5 percent to 40 percent, but from what I have learned, the loss rate was probably near the higher end of that spectrum. With every one of my father's combat missions, he saw loss of life, injury of colleagues, and loss of aircraft. At the same time, he had to stay focused and fly while coordinating navigation, communication, bombing, and approaches of enemy aircraft. In addition, he had to wake up extremely early in the morning, dress for combat, be briefed, and climb into his ship to face possible death once more. It would have been mentally taxing to experience those horrific events, especially for a young man in his early twenties.

On that fourth raid, Ben witnessed an event that he would not forget:

I saw a most sickening sight right after turning off the target. A B-17 in the formation ahead of us caught fire in #3 engine. It veered off to our right, rolled over on its back in a dive, completed the roll, and dove straight down. The wing came off just as I lost sight of it. Another Queen died proudly.

There are no words to describe how men felt as they witnessed death every day, not knowing if they would be next. However, they bravely went on their next missions out of a sense of duty to their country and their fellow airmen. They did it for the freedom of their loved ones back home, although they longed for their twenty-five missions to be completed so they could end their time at war.

Ben's Fifth Combat Mission
July 17, 1943

Although the 92nd BG had experienced a tough combat mission on July 14, 1943, there was evidently a pressing need for some practice formation flying. Timing and accurate formation flying were a must for effective air war posture. In fact, it was said that the Luftwaffe pilots targeted sloppy formations because they assumed that the pilots were not well trained. Flying out of formation opened the bombers to enemy fighter attack, frustrated bombing accuracy, and resulted in some planes being hit by bombs dropped from planes in their bomber stream from above. The emphasis in practice was always to fly in a tight formation. Wing tip to wing tip was the way they were to fly. That was nearly impossible, and it took all their physical strength, especially on long missions, to maintain that position. With the unpredictable and unfavorable weather, if the formation was not tight and precise, some of the aircraft collided.

When I visited the area around Alconbury in Cambridgeshire, England, I was surprised by the proximity of the WWII air bases to each other. They were just a few miles apart! Re-emphasizing the need to fly tightly, the 327th BS, on July 16, was devoted to practicing forming up and flying in formation. My father recalled:

We had a practice formation flight today, and I flew first pilot for a change! I should get another crew of my own soon and another ship of my own.

During that time, replacement planes were being manufactured at astonishing rates and sent to the European Theater. Likewise, due to heavy losses of aircrews, men were being trained and sent to Europe to supplement the deficient ranks of fliers to accommodate the growing air war. As my father's YB-40 had been badly damaged in his first combat mission, he and other members of his original crew flew on other aircraft with other crews.

On July 17, Ben was assigned to the mission to Hannover, a large German city that would be the target for eighty-eight air raids by the RAF and the USAAF during the war. Hannover was an important railway junction at the intersection of two major east-west and north-south routes. It was also the fifth most active industrial center in the Third Reich, producing tires for military vehicles and aircraft, other rubber parts and products, guns, and batteries for submarines and torpedoes.

Raids on Hannover involved a relatively short flying time from bases in the United Kingdom, and the nearby Lake Steinhuder Meer was a useful navigational aid. Other targets for the raids on July 17 were the Fokker aircraft plant in Amsterdam and a shipyard in Hamburg, with a diversionary target for B-26s into northeast France.

Lieutenant Colonel William Buck and Captain Donald Parker led the CW for the 92nd BG with sixteen B-17s and two YB-40s departing at 07:18 a.m. No wonder they had been practicing formation flying a day earlier! The night before and during the early morning hours of July 17, RAF Allied air forces attacked enemy operational bases in Belgium, Holland, and northwest Germany.

There were four bomb groups of B-17s on July 17 headed for three targets. It is interesting to note that, in addition to the usual five-hundred-pound bombs, the bombs loaded for that mission included British 250-pound incendiary bombs and A-1 incendiary bombs used to set the target ablaze.

The plan for the day was to attack Hamburg with one force of bombers and attack Amsterdam with another at about the same time. These were to be followed by an attack on Hannover by two bomber forces. The plan was intended to confuse the enemy fighters in the area, thereby reducing the strength of the fighter opposition that could be brought to bear on any one unit.

Even though the raid to Hannover was called back due to cloud cover, there were targets of opportunity, and thirty-three B-17s dropped bombs over northwest Germany. Of the sixteen B-17s dispatched by the 92nd BG, all were recalled, and no bombs were dropped in enemy territory. The bombs that were not used were jettisoned, meaning they were discarded to reduce the plane's weight, lower fuel consumption, and mitigate the explosion risk if the plane was hit on the way home.

My father described that day in his journal:

I went on my fifth raid today, thereby becoming eligible for the Air Medal. There was an overcast over the target area, so we were called back. The German fighters, Me-109s and FW-190s, jumped us before we reached the enemy coast and stayed with us even after we were within sight of England. We only had twenty-three ships in our formation, and we were subject to all of the enemy attacks. We did violent evasive action, and I think that is why we all returned. We had eight men wounded, and most of our ships were damaged. Our own ship got two hits in the leading edge of the left wing, and two .303 bullets went into our nose, slightly wounding our bombardier-gunner. It was a long running fight, and the fighters were coming in fast and close—all from the nose. The radio news broadcast says fifty enemy fighters were destroyed, for two Fortresses lost, but that includes all operations today by B-17s stationed around here. We ourselves lost no ships and destroyed about five fighters.

I wish that I had a picture of my father and his crew in front of one of his B-17s or YB-40s. As the YB-40 was a secret airplane, I understand why there were no pictures of him with it. However, there are many pictures of the complete ten-man crews in front of their B-17s. Unfortunately, the only picture that I have of my father when he was in Europe is Figure 4-6, which shows him sleeping on a cot in a tent. With little rest and the rigors of combat, it is no wonder that he took a cat nap, even under arduous conditions. When he earned his Air Medal after five combat missions, he had only twenty more combat missions to go before he could go home. He was fatigued and war-weary but ready to fight again.

Ben's Sixth Combat Mission
July 24, 1943

For the first time in the European air war, Allied B-17s were sent to occupied Norway to attack three targets. On July 24, 309 B-17s were dispatched on the longest-ever mission undertaken by B-17s of the 8th BC. The long mission was unprecedented, as it was a 1,900-mile round trip to the edge of the Arctic Circle. The targets were magnesium, aluminum, and nitrate plants at Herøya, Norway, and U-boat workshops and harbor installations at Trondheim, Norway.

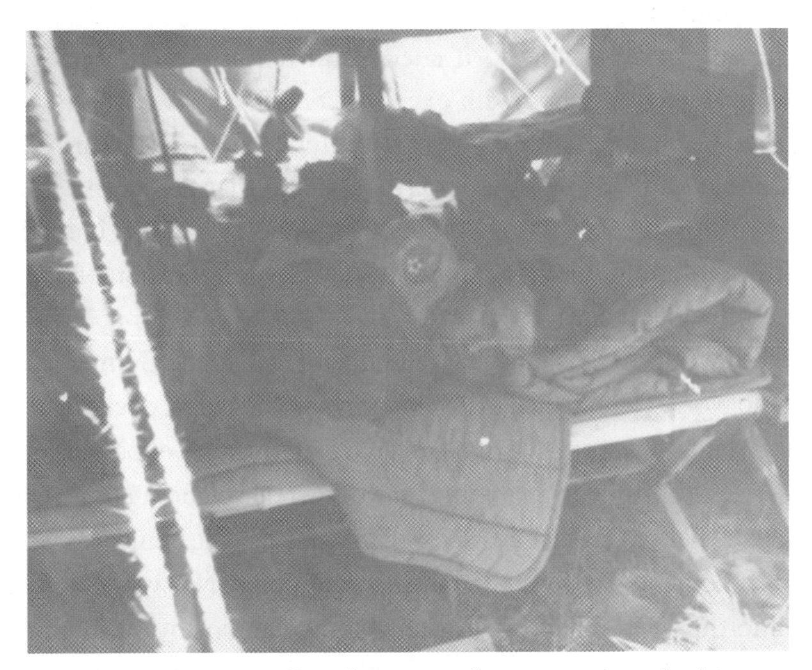

Figure 4-6: Ben Smotherman taking a nap amidst his combat missions, Alconbury, England, 1943

Following the occupation of Norway by Nazi Germany in April 1940, the German occupying force devised plans to build the aluminum and magnesium plants in Herøya. However, the planned construction of the plants ended when the bombing by the B-17s completely destroyed existing facilities there as well as parts of the surrounding areas. A French observer, Colonel Corniglion-Molinier, said that when the bombs hit Herøya, the entire factory there went "poof" after the American airmen planted it with their precision bombs. He called the raid, "wonderful, marvelous, beautiful."

The target at the Trondheim port was also briefed as the heavy water plant at Rjukan, Norway. On the eve of World War II, scientists both in Germany and Great Britain realized that heavy water could be used to make nuclear weapons. Trondheim was the big German U-boat base, and the approach to the base was sheltered by the fjords of Norway.

It was home to the Nazi warships and submarines that preyed on the northern supply route to Murmansk. It was also the last reported haven of Germany's great battleship, the Tirpitz, and the ten-thousand-ton cruisers of the Admiral Hipper class.

The mission was deemed successful as targets at Herøya and Trondheim were bombed with excellent results by 208 aircraft. From the 92nd BG, fifteen aircraft were dispatched, including one YB-40. An Associated Press newspaper article stated that the strong force of American Flying Fortresses pounded Trondheim and left it a "raging mass of exploding bombs, flaming oil tanks, and black smoke, which mushroomed up thousands of feet."

Photographs of the damage at Trondheim, called the German "Gibraltar of the North," confirmed that submarine repair shops, docks, and other naval installations were heavily damaged by thousands of pounds of high explosives according to an intelligence officer. He elaborated, "We really hit it on the button, for pictures showed that not a single bomb was wasted."

This momentous occasion of the first mission to Norway took months of extraordinary training by the aircrews and planning by the intelligence community. The pilots, navigators, and bombardiers had to learn about the geography of Norway, its potential targets, the ways to navigate to a new country, the flak gun locations, and the likely fighter coverage by the Axis countries.

My father was very excited to go on this historic and noteworthy mission. He wrote in his journal on July 24, 1943:

We went to Norway today—my sixth raid—and it was a very long trip—over eight hours. Our ship was leading the wing, and I flew as tail gunner to observe the formation for the colonel. It was an interesting experience. We had a lot of accurate flak over the target but didn't see many fighters. All of our ships returned safely.

Colonel William M. Reid was flying as the Air Commander, and Captain Roland L. Sargent from the 327th BS led the 8th BC strike. The 92nd BG could not get to the target because of the weather about fifty miles before the initial point where the bombers would turn and commence the bombing run, so the box formation swung around and came in from the northeast.

As my father's ship was leading the CW, his navigator was navigating for the entire formation and had quite a challenge in finding an alternative path. They overflew their destination, which was clear of clouds directly overhead. From there, Colonel Reid executed a 180-degree left turn and approached from the northeast with a clear view of the target. Bombs were then dropped manually during a ninety-second bomb run. For Colonel Reid's leadership in setting up a new bomb run after weather had prevented the planned approach, he was awarded the Silver Star.

As my father was in the tail gunner position as far aft as a crewman could be in the aircraft, in a very cramped position, he had an excellent view of the formation in flight. He also observed how a well-executed mission coordinated the efforts of the pilot, copilot, bombardier, and navigator. All went well, except their waist gunner, Staff Sergeant Curtis L. Moore, had a slight leg wound.

It is interesting to read how the various bomb groups responded to the new and distant target. Some of them missed the Norwegian initial point by several miles and followed the coast to find the target, and some overran it on the way there but managed to find it on the way back.

The raid caught the enemy off guard, so the opposition was weak. It was debriefed by the aircrews that the majority of the German pursuit planes seemed reluctant to attack. Only four of the nine BGs reported any enemy encounters, and those were five to fifteen aircraft (mostly FW-190s and Me-109s).

The hot flak shrapnel was moderate, but it did damage one B-17 enough that it was forced to land at Vännacka, Sweden. It was the first aircraft and crew ever to be interned in Sweden during WWII.

This was the longest mission yet for the 8th BC and the first mission to Norway. All aircraft were equipped with long-range tanks and flew at low altitude until just off the Norway coast to conserve fuel. As an example of the type of maps shown in the 92nd BG reports, the complex routes flown to Herøya are depicted in Figure 4-7.

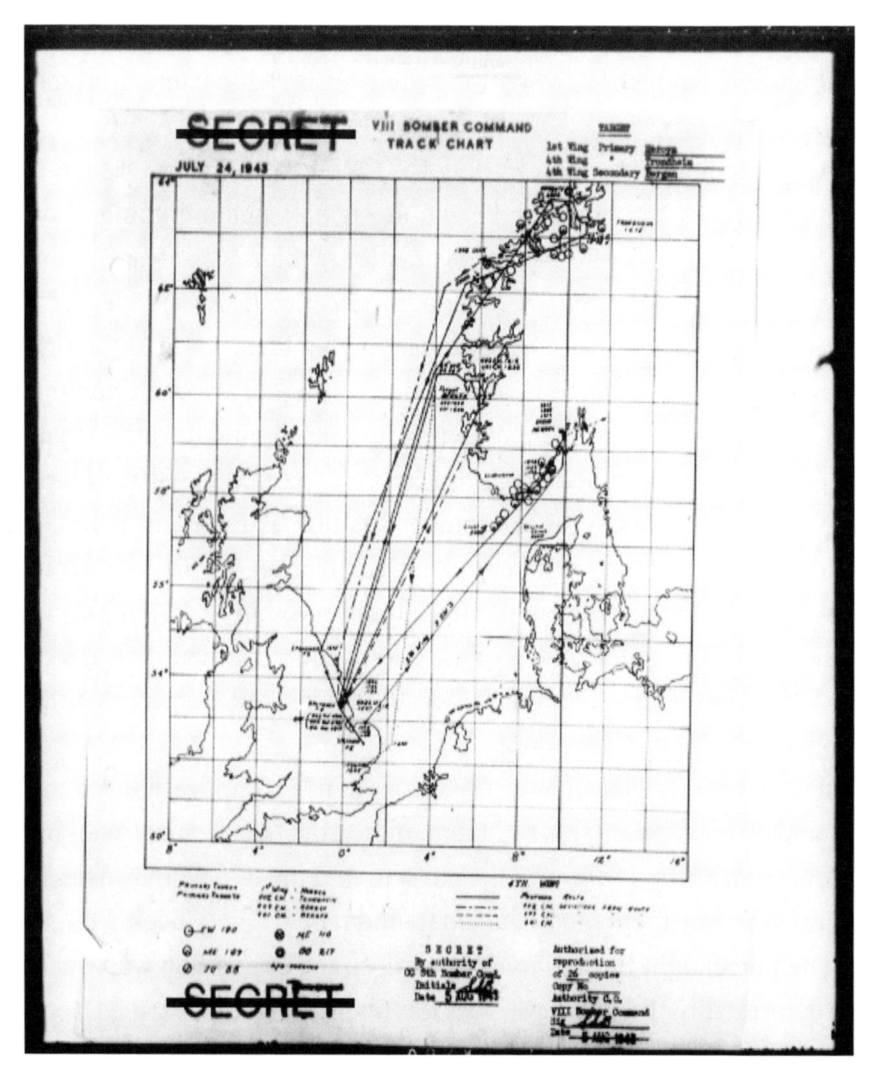

Figure 4-7: Map of routes to Heroya, Norway for 8th BC bombing mission #75 on July 24, 1943 (Air Force Historical Research Agency Microfilm Reel #A5933)

In my father's last letter home, dated July 27, 1943, he told his family about the raid, saying "I haven't been demoted but went as tail gunner in order to observe the formation, as we were leading. It was an interesting experience, and I had a lot of fun! I'll try to write you a letter within the next few days… Don't worry when you don't hear from me, for I get sorta busy at times! So be good now and take care of yourselves." He always signed his letters: "Love to all, Ben."

Final Days in England
July 26 – 27, 1943

Ben Smotherman would not fly again until his last mission on July 28. However, he had two more entries in his journal before that fateful day. He wrote:

July 26, 1943

Bill is missing tonight. I still have hope that he is alright, but there is nothing official yet to put us at ease. We're all anxiously awaiting word that he and the crew are safe and praying that all is well. Casey and Baker were with him, and Cates went down with another crew—all of them came over with the YB-40 bunch. I somehow feel that Bill and the others are okay and certainly hope and pray that they are.

I didn't go on the raid today. My crew was alerted but was scratched at the last minute last night.

The mission for the 92nd BG on July 26, was to Hannover. Seventeen B-17s and two YB-40s were dispatched, with eleven B-17s and one YB-40 completing the mission.

Two B-17s in the 92nd BG were ditched in the water off the English coast on that mission. One of them was Hell-Lena (42-29981) piloted by my father's buddy, Lieutenant Paul S. Casey. That crew also included my father's best friend, Bill Stewart, and navigator Luther Howard Baker. No one was injured, and they were all picked up from their dinghies by the Royal Forces at night after four and a half hours. Figure 4-8 shows the dramatic photograph of the downed aircraft and the crew in their dinghies.

However, it wasn't a happy ending for everyone. A 92nd BG B-17 called "Yo Brother" (42-29709), piloted by First Lieutenant Alan E. Hermance, tragically crashed into the North Sea about ten minutes off the island of Norderney with its tail chewed up. All ten aboard, including my father's colleague Velgene V. Cates, who was the copilot, were killed.

Figure 4-8: B-17 (42-29981) and its crew (including Ben Smotherman's friend Bill Stewart) in a life raft after ditching in the North Sea on July 26, 1943, taken by reconnaissance aircraft from Coastal Command, RAF Bircham Newton (photo courtesy of American Air Museum, IWM Duxford)

July 27, 1943

Bill and all the crew are safe. We heard about it when we got up this morning, and there were a lot of sighs of relief when we got the news. They went down in the Channel and were picked up last night. They had quite an experience and have quite a story to tell!

That was the last entry in my father's wartime journal that he had been writing to my mother. I am glad he found out that some of his friends were okay, and he probably did not know about his other friend who was KIA. I hope that he was able to hear those stories around the bar the evening of July 27, as that would be his last night in Alconbury.

Chapter Five
Shot Down

My uncle, Don Smotherman, Ben's brother, was in training to be a pilot and lamented to his family how disappointed he was that he was not in combat. Ben's best friend and roommate, Bill Stewart, rebuked Don in a letter written in September 1944: "I hope like hell you will never see combat, especially in this theatre (Europe). I know how you feel, that you have gotta fight and be a big brave hero, but boy, let me tell you this war ain't it! I've seen experiences that shouldn't happen to human beings. You never know whether you will get back once you cross that [English] channel."

I believe that strong admonishment from the man who was my father's roommate through cadet training and while overseas captured the general feeling of the airmen who were rousted out of bed to perform their duty—another combat mission. It was not a duty the young men wanted or liked but one they were committed to.

For that fateful mission to the Fieseler German aircraft plant, which made airplane parts for the FW-190 at Kassel, Germany, 180 B-17s and two YB-40s were dispatched. My father was in the left seat for his seventh combat mission as the first pilot, or "pilot in control," in B-17 42-29798. In his own words, he described the mission in his POW Wartime Log:

It was July 28, 1943. We were awakened in the wee hours of the morning [2 a.m.] *and went to eat breakfast—with no eggs! We were then briefed for our target, which was Kassel, Germany. It looked like a rough one! I was glad that I had a good crew for that one.*

When I arrived at the ship, things didn't look so good. There were several things which weren't as they should have been. The ground crew worked hard, and we finally were ready, after a fashion, just in time to taxi [Figure 5-1].

Figure 5-1: Preflight repairs, July 28, 1943, from BFS Wartime Log

We dropped into our position and taxied out for takeoff. In position at the end of the runway, we awaited the zero hour. Then, with a sudden roar, the lead ship moved down the runway, followed by a second ship, and a third. One by one, the ships moved out and raced down the runway and rose into the air. Our turn came, and we moved into position. Everything was all set. We were off!

Leaving the runway, we veered to the left, cutting off the ship ahead of us. In a few minutes, we were in formation, circling the field. Heading out on course, we made our way to our rendezvous. Soon, we could make out other formations, which were joined by still others until the sky was literally filled with airplanes! It's a thrilling sight to look about—above, below, on all sides—and see hundreds of airplanes in formation on a common course.

Crossing the English Channel, the crew checked their guns. We were ready for the fight we knew was coming! Soon, we could see the coast of Europe far below us (we were flying at 25,000 feet) and our hearts beat a little faster. Soon after crossing the coast, our protective fighter escort left us, and we were on our own.

Then, the flak started. There is little or no defense against flak, so all we could do was watch the black bursts and maintain our position in the formation. I flew the ship, and my copilot handled the throttles, and we didn't have time for much else!

As we neared the target, the German Luftwaffe fighter attacks started—FW-190s. Their mission was to disrupt our bomb run [Figure 5-2].

On the bomb run, we were vulnerable to fighter attacks because we had to fly straight and level. Yet one could always tell when a fighter was aiming at his ship, but that was small consolation! Bombs away [Figure 5-3]*!*

We held formation and watched the bombs stream out of the other ships. Bomb bay doors were closed, and we turned off the target run and set a course for England. This course was planned to avoid areas defended by flak guns.

Figure 5-2: Flak and the German Luftwaffe attacks, July 28, 1943, from BFS Wartime Log

Figure 5-3: The bombing run on Kassel, Germany, July 28, 1943, from BFS Wartime Log

Due to weather, several aircraft abandoned this mission, and only forty-nine made it to the target in Kassel and bombed it. Bombing occurred from about 10:30 a.m. to 11 a.m. on that Wednesday in July. The bombers battered the aircraft factory at Kassel and destroyed more than sixty Nazi fighters during battles in which twenty-three heavy bombers and one Allied fighter were lost.

There were a few firsts about the mission. It was the first time the P-47s flew with extended-range fuel tanks attached to their bellies that could be jettisoned after use. The extra fuel extended the range of the P-47s by about thirty miles, thereby increasing their escort support for the Flying Fortresses. Due to the extended fuel tanks, they were also able to attack more enemy aircraft, who were surprised to see them!

However, the Germans had surprises of their own. They were using several rocket-propelled bombs, which were highly accurate. In the spring of 1943, the Luftwaffe had begun equipping some of their fighter planes with a rocket developed by the German Army. The tube-launched, spin-stabilized rocket was believed to be larger than forty millimeters, and large flames came from these rocket guns. The bursts were almost as large as flak bursts. The rockets were launched from a cumbersome launch tube mounted beneath the wings of FW-190s and Me110s. The attack technique enabled enemy aircraft to get 1,500 to 2,500 yards away from the B-17 formation, raise their nose as they approached, and lob their shells into the bomber. Apparently, the enemy aircraft were not capable of continuously firing the rockets, so they had to break off after firing one or two shots.

Enemy opposition was intense and aggressive during the mission. Over 250 single-engine fighters were spotted. As the bombers retreated from Kassel and took a path over Holland to get back to England, the German fighters were on the attack until 12:30 p.m., at which point the Allied Spitfires and Thunderbolts escorted the bombers home. Unfortunately, my father's plane was one of those pursued by the German fighters.

My father's friend Bill Stewart was not on the raid but explained in a letter to Ben's brother Don: "Ben was pilot of the ship. It was his first raid as pilot. He had a bad position to fly, tail end Charlie, as we call it. The German fighters like to pick on tail end Charlie. I did not go on the raid, so I didn't see him when he got hit. I understand an FW-190 got his

#4 engine, and he started dropping back, and then the fighters really come in when you lag behind [the formation]. He was last seen going down under control."

As my father later told the story:

Suddenly, the number four engine "ran away" (went to high speed). We later learned that a fighter far behind us had hit us with a rocket—fairly new in those days. We went through our procedure to feather the propellor, but it would not feather. Looking out, we could see the engine covered with oil [Figure 5-4].

Figure 5-4: Damaged engine covered with oil, July, 28, 1943, from BFS Wartime Log

The drag of the inoperative engine slowed our speed, and we were unable to hold our position in the high squadron. The formation was descending to a lower altitude. I maneuvered into position in the low squadron, hoping I could maintain that position. No luck! To make matters worse, my number three engine was not putting out full power, and I suspect it had been damaged when number four was hit. As our formation pulled away, I tried to join a formation behind us, but it too pulled away. Then, we were alone, and I knew what was coming. I looked for a cloud bank to hide in, but the sky was absolutely clear! Then, the fighter attacks started—FW-190s.

As a flying cadet, I had trained as a fighter pilot, so I put that knowledge, along with everything else I could remember, to work in my efforts to avoid getting hit. Fortunately, the fighter pilots employed the classic fighter attack pattern of flying parallel to us until they were ahead of us, then turning into us for a nose attack—the favorite pass of the Luftwaffe pilots. As they turned into us, I turned into them, which spoiled their run. As they got within firing range, I would skid sideways or jump the airplane up and down until I feared I might hurt one of my crew!

The FW-190s were soon replaced by Me-109s, and we knew they were calling ahead and scrambling fresh airplanes to try to shoot us down.

So engrossed was I at one point that my copilot pointed to the compass and informed me that I was heading back into Germany [Figure 5-5].

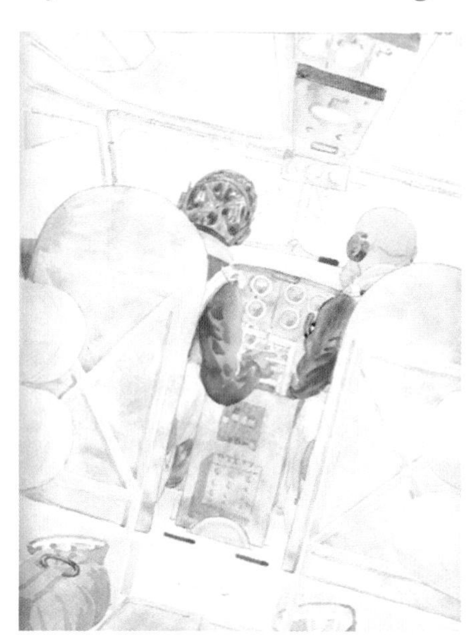

Figure 5-5: Inside the cockpit after being attacked by the Germans, July 28, 1943, from BFS Wartime Log

During the violent maneuvering which I had to do, we continued to lose altitude. During a rare, brief lull in the fighting, I briefed my crew over the intercom. As soon as we were over the channel, the engineer was to transfer fuel to the three operating engines, for we were running dangerously low on fuel. The navigator was to give me a heading to take us directly to the English coast. The radio operator was to contact ground control for a vector to the nearest airfield. But we didn't get a chance to do all of that.

We were now "on the deck" (flying about five hundred feet above the ground), and all the gunners were out of ammunition. Then, a surprising thing happened: Two of the Me-109 fighters dropped their landing gear and flaps and flew along beside us [Figure 5-6]*! From what I had been told, that was a signal that they would escort us to the coast and let us go home. Our spirits rose! But that was not to be because they pulled away and started attacking us from the rear. I could see the tracers going by us. I had been hit in the wing on a previous mission, and I knew we would be temporarily out of control with no altitude to recover. We were practically out of gas because we had been under attack for almost an hour. To make matters worse, I saw towns ahead, and I didn't want to kill any of the people of Holland. I told the crew to prepare for a crash landing. I turned back toward an open field we had just passed, and I landed in that field. I think my tail wheel caught the wire fence as we crossed the edge of the field* [Figure 5-7].

Figure 5-6: Luftwaffe pilot flying alongside Ben Smotherman's B-17, July 28, 1943, from BFS Wartime Log

Figure 5-7: Ben Smotherman crash landing his B-17 in Holland, July 28, 1943, from BFS Wartime Log

When the ship skidded to a halt, we were all out of the ship in a very short time. None of the crew was hurt. We sustained no hits on the ship during the fight. Now we had to get back to England!

That afternoon of terror, adrenaline, and fighting the German attacks resulted in the crippled B-17 landing at the Binnenveld area near Wageningen, Holland, about noon, as shown in Figure 5-8. For my father's efforts to turn around and not hit any of the Dutch towns, and for bringing his crew down safely, he would later be awarded the Distinguished Flying Cross.

Figure 5-8: Ben Smotherman's crashed plane B-17F (42-29798), July 28, 1943 (photo courtesy of Peter den Tek, Vliegeniersmuseum, Netherlands)

After landing, I assume that they tried to destroy the Norden bombsight as they had been instructed to do since that technology was top secret, and the Germans really wanted it. Equipped with only some money, a map, and a compass in their escape packets, the tired and hungry crew left their ship. With knowledge of the underground that would help the downed Allied forces, my father and his crew tried to escape from the Germans who were in hot pursuit. My father continued his agonizing story:

I instructed the crew to split up in pairs and to hide out somewhere. I tried to set the airplane on fire but was unable to do so [Figure 5-9]. *The German fighters were circling overhead. My copilot and I swam a canal to get to a small group of Dutch farmers who offered us clothes. We declined because we knew they would be killed if they were caught helping us. We made our way to the middle of a wheat field and waited there for darkness.*

Figure 5-9: Ben Smotherman trying to catch his downed aircraft on fire, July 28, 1943, from BFS Wartime Log

Just before darkness, we left the wheat field and started walking down the road. We found a road sign, which read "Arnhem 26 kilometers." We had been briefed on how to contact the underground in Arnhem, so we decided to try to make it there that night. We passed German soldiers on bicycles who paid us no heed. We knew there was a curfew in Holland, so after dark, or after 11:00 p.m., we would hide by the side of the road whenever a vehicle approached. Suddenly, we found ourselves at the edge of a village, and we were trying to decide whether to walk through the village or try to make our way around it when we heard a vehicle approaching. We went in between two buildings to wait for it to pass. When we started to come out, there was a man there with a pistol. I started to escape through the back of the buildings, but they came together in a "V" with only a closet at the juncture. I walked out and surrendered. The man took us into a tavern (where he had been sitting when he spotted us) and offered us some Schnapps. We declined and asked for water since we had not had a drink all day long! He gave us a drink of water, and we waited there for the Luftwaffe to come and get us. A Luftwaffe officer in a white jacket took charge of us, and we were put in a truck to be taken to jail.

While we were in the jail, one of the German pilots who had fought with us came to visit us. Since my copilot spoke French, he talked with him. He told us that we had shot down seven of their fighters during the fight!

Apparently, the commander of the Dutch air raid service had notified the deputy of the Wageningen police department that an American bomber had crashed in the vicinity of the "Rijnsteeg/Veensteeg," and two police officers had immediately gone to the emergency landing site. By then, the crew of ten had already fled.

Around 1:00 p.m., the police officers apprehended four of the crew-members and transferred them to the commander of the Wehrmacht (German armed forces), who were also at the scene. They requested that several roadblocks be created and that the plane be guarded, and they specified they would need help finding the rest of the crew.

At 11:20 p.m., the commander of the Dutch Police brought my father and his copilot, David Pollak, to the police station where they were jailed. They were transferred to officers of the German Wehrmacht at twenty minutes past midnight.

My hope is that the six airmen were fed and able to get some sleep after a long and frightful day. They were undoubtedly scared, not know-ing what the next day would bring. Although they had been briefed about what to say and do as POWs, the reality of the situation weighed heavily on the entire crew. *What will happen next? Where will they take us? How will we be treated?*

The next morning, July 29, at 10:15 a.m., two police officers, with the help of local workmen, found two more of the downed aircraft's crew—waist gunner, Fred J. Rentz, Jr., and tail gunner, James W. O'Connell—in the town of Ede. They were brought to the station at 11:30 a.m. At some point, the remaining two crewmembers were also found and turned in to the Wehrmacht. The names of the crewmembers not already mentioned were radio operator, Jack Fogarty; waist gunner, Curtis L. Moore; bombar-dier, W.V. McKenzie; navigator, Sidney J. Leigh; engineer, J. Atlee Pear-son; and ball turret gunner, U.M. Chandler.

As my father told it:

We were taken from the jail the next day and put on a train. We spent the next night in the interrogation center, as I recall. From then on, it was travel, interrogation, and travel until we finally wound up at the Prisoner of War camp at Sagan, Germany [now Zagan, Poland], *Stalag Luft III.*

Meanwhile, at Alconbury, the officers, ground crews, and flight crews who had not been on the Kassel mission on July 28 watched the B-17s and YB-40 return to base and noted the planes and aircrews who did not return. There would have been great sorrow for the ground crew of my father's B-17 to not have their ship return and sadness throughout the squadron with the loss of crews who did not make it home that day. There would have been reports of 42-29798 being hit and falling out of formation, though there were no witnesses to the crash landing. So, no one yet knew whether the crew was okay or what had happened to them.

Also, on the evening of the 28th, the operations people in the 327th BS were instructed to pack up the belongings of the ten missing airmen. Each had a footlocker, and all personal belongings were packed up in the footlockers, which were sent back to the families in the United States. From there, the vacant beds would be freshly made, and ten new slots would be available to house a replacement crew. The changeover was immediate.

Sadly, there were no more entries in my father's war journal, and my mother's scrapbooking of the war ended. I remember her telling me that she did not know whether he was dead or alive, only MIA.

On August 4, 1943 (a full week after he was shot down on July 28), a Western Union telegram was delivered by hand to my mother. It stated, "I REGRET TO INFORM YOU THAT THE COMMANDING GENERAL EUROPEAN AREA REPORTS YOUR HUSBAND FIRST LIEUTENANT BENJAMIN F SMOTHERMAN MISSING IN ACTION SINCE TWENTY EIGHT JULY OVER KASSEL GERMANY IF FURTHER DETAILS OR OTHER INFORMATION OF HIS STATUE [sic] ARE RECEIVED YOU WILL BE PROMPTLY NOTIFIED [James] ULIO THE ADJUTANT GENERAL." Then, there were newspaper articles about his last raid and the fact that he was missing. My mother saved these articles as last remnants of Daddy Ben flying in the war.

My mother's parents, my father's family, their friends and acquaintances, other wives of officers overseas, and the clergy did all that they could to help the grieving and distraught young wife who waited to find out if she was a widow.

Chapter Six
For You, the War Is Over

My mother and Ben's family were devastated that Ben Smotherman was MIA, and there was a community effort to allay their fears that he had been KIA. For example, Bill Stewart wrote to my mother on August 6: "Dearest Dolly, This letter I have put off because I have been quite upset over last week's experiences. I am sending these money orders to you that Ben wanted me to send just in case he didn't get back right away. He will be back but had some bad luck, probably escaping or either a prisoner of war now… I have Ben's things. Not his clothes because I packed them and they will keep them until he gets back. I kept his personal things. We agreed between us what things we were to keep especially. I don't want you to let this happening get you down too much, Dolly. I am sure Ben's okay wherever he is at this moment. All the boys here feel the same way I do… He had a chance to get away I know. Don't you worry because he will pop up one of these days and have a big story to tell all of us… Things have been rough as hell this past week or so and seem to be getting tougher. Let's all hope it will end before many months."

Another friend, navigator Joe Liebman, further elaborates in a letter to my mother dated August 21, 1943: "I know that it must have been a terrible shock to you when you learned that Ben was missing in action, as it was to me after having returned from the same mission. I know how your heart is filled with anxiety and you're all mixed up and very fearful. I did not see Ben's ship leave the formation, but on good authority from one that did, I can tell you as much as was seen… Because of apparent trouble, Ben was forced to lag behind, and when last seen, the ship was under complete control. So, I would say that there is a very good chance that Ben is now on the continent of Europe, most probably a prisoner of war… I do not need to tell you what I and all of the fellows thought and still think of Ben. He's a grand fellow and very capable of taking care of himself. That is why we all feel that he is safe and well. Above all, do not worry, as we are all confident that Ben is fine."

It strikes me how mature and kind these friends of Ben were. I think the tragedy of war and the courage needed to face war every day matured these young men who were in their early twenties. They had lost a friend and were uncertain of his future. It was distressing for them to wonder whether their comrades were dead or alive. Yet they took the time to write letters to my mother to give her hope and to tell her what they had witnessed firsthand concerning his plane.

The letters also revealed that each crewmember thought about what should be done with his personal effects if he did not return. The airmen were faced with the reality of that every day—they may not make it back, and they needed to decide ahead of time what should be done.

Of course, receiving the MIA telegram was heart wrenching for my mother and Ben's family. However, they received significant support from their community. For example, just like Bill and Joe, many friends, relatives, wives of fellow airmen, and clergy wrote to Ben's family and to my mother.

I was moved by the perspective in the letter that Bill Stewart wrote to Ben's sister, Jack Darby, dated August 26:

"I can assure you I did gain something by being Ben's friend. You know Ben and I went through cadets together and have been together all through our training. I can assure you, since he has been missing, I have been lost without him. We were always together just like a brother, I will say.

"When he didn't come back that day, I went about finding out all I could from different boys who went. When Ben was last seen, his ship was under control… I cannot swear to these statements because I didn't see him. I feel deep down that Ben is okay… This war is <u>hell</u>! I mean just that. You never know who is going to be next, yourself, your friend, or who. In case you don't know, that was Ben's seventh raid. So far, I have made eleven raids. If and when I get twenty-five [missions], I will get my chance to go back.

"I was shaken up pretty bad [from his plane ditching], and then when Ben didn't come back the next day, well, you can imagine how it got me. A month has passed now, and I feel a lot braver. Many of Ben's and my friends have gone down this month. This is a terrible thing, and the sights I have seen are not at all encouraging. I keep hoping and praying that this thing will be over before many months now...

"My chances of coming out of this thing aren't good, like the rest of us boys. All I ask is that I get a chance to bail out and save my crew. I don't care about all the ribbons and medals you receive. They don't mean a thing. Being alive after it is all over is all that counts.

"I want you to know that this business is not at all safe. I feel like this—that if I get through, I will be the luckiest man alive. These German pilots are good fighters, and they mean business. They are not whipped by a long shot. We will lose many planes before they are. They are growing weaker, but they are far from weak."

Unbeknownst to Ben's family and the families of his crew, from the jail in Holland, the aircrew of 42-29897 traveled by train over the next three days toward Frankfurt am Main. At the train station in Frankfurt, a white bus with green crosses labeled Dulag Luft picked them up. A short drive later, the captured airmen arrived in Oberursel at Dulag Luft, the Luftwaffe interrogation center.

The captured prisoners had been briefed, drilled, and trained at their bases in England on how to respond to the masterful interrogation by the Germans. Without a doubt, all on the bus to Dulag Luft (short for *Durchgangslager für Luftwaffe* – transit camp for the Air Force) were reminding themselves of these briefings while still in shock and weary from the long few days after being captured. They were scared, but they were brave.

One of the first things that the prisoners saw was "POW" outlined in white rocks at the entrance and painted on the roofs (Figure 6-1). It was a desolate place surrounded by barbed wire but had no guard towers or perimeter floodlights. Sitting at the foot of the beautiful Taunus Mountains, it was where all captured aircrew were held for interrogation before being transferred to their permanent prison camp.

Dulag Luft had several parts: the interrogation center, the "cooler" comprising solitary confinement cells for men awaiting interrogation, and the barracks at Oberursel, as well as the hospital at Hohemark for seriously injured prisoners.

Figure 6-1: Dulag Luft interrogation camp (photo courtesy of alchetron.com)

During the first nine months of 1943, when my father arrived at Dulag Luft, there was an average monthly intake of one thousand prisoners, half British and half American. As solitary confinement was the rule, the capacity of the camp was supposedly limited to two hundred. Yet, on average, there were 250 prisoners imprisoned there per day during the course of the war. As more Americans were shot down and numbers grew at Dulag Luft, interrogations became shorter and men were shipped out faster. Later in the war, due to the bombing of Dulag Luft, the interrogation center was moved to Frankfurt am Main and then to nearby Wetzlar.

The camp at Oberursel had four long wooden barracks. Two of the barracks with two hundred cells each were connected by a passage and constituted the "cooler." The third barrack contained the administrative headquarters, and the fourth barrack, an L-shaped structure, held the interrogation offices, files, and records. A picture of the barracks and "cooler" is shown in Figure 6-2.

Within its buildings, Dulag Luft had every known device for extracting information from the captured airmen, and the Nazis knew all the tricks to get information from unsuspecting POWs, including using hidden microphones behind the wooden paneling. Unfortunately, I don't have any journals, records, or discussions by my father about what the interrogation process at Dulag Luft was like, so I have gathered information from other POWs who went through interrogation there.

Figure 6-2: Dulag Luft barracks and "cooler" (photo courtesy of Clark Special Collections Branch of the USAF Academy Library)

The German procedure was that all captured airmen were taken to Dulag Luft with any documents or paperwork retrieved from their downed aircraft. These were turned over with the airmen's dog tags. The men's personal belongings were each carefully catalogued by the German intake personnel to be returned. Most were, but sometimes leather A-2 jackets and wrist watches issued by the military to officers were not.

The initial comments made by the German officer to the prisoner upon first meeting him were: "Vas Du Das Krieg Est Uber – For you, the war is over. You are now a prisoner of the German government for the remainder of the war." From then on, my father, like the other USAAF downed airmen, would be known as a "Kriegie," short for the German word *Kriegsgefangener*, meaning prisoner of war.

Upon his arrival, each captive airman went straight to the reception center building where he was stripped and searched and had to turn over his personal possessions. This is also when the aircrew were separated from one another for the first time. Each prisoner would be allocated to a reception room after being escorted to the washrooms. They would be permitted to use the toilet but only under strict supervision and were able to use a dull razor and a few toiletries of questionable quality. In addition, the prisoners were denied cigarettes and Red Cross food.

After procedural matters were taken care of, the prisoner would be isolated in the cooler. Each cell was eight feet high, five feet wide, and twelve feet long and held a cot covered with hay, a small table, and a wooden stool. There was also a metal handle or bell outside the cell's heavy door to call the guard to signal the prisoner needed to go to the bathroom. Such calls were seldom responded to quickly. The more information given, the more attentive the guard was to the prisoner's needs.

A small transom window was high on one wall but was frosted over. According to Kenneth W. Simmons's memoir, *Kriegie*, a brief respite was to go to the toilet down the hall, wash up, and experience light and a view through a window. This seemed to relax and refresh prisoners, and Simmons stalled every time he was using the toilet with the guard standing outside the door!

Every aspect of confinement was designed to create psychological pressure, thus breaking down any resistance to interrogation. Rooms were kept in dim light for extended periods of time. Some prisoners climbed to look out the transom window. In August, when my father was there, it was already hot. Periodically, heat was turned up in the cells to make it more uncomfortable for the captured prisoners.

Many hours after a prisoner was put in a cell, a German saying he was representing the International Red Cross would arrive, bearing a bogus form. He would say that once it was filled out, the prisoner's family could be notified that he was alive and well. The form questions started out fairly innocuous but soon became more specific to military information.

Per the Geneva Conventions, the only information that a prisoner was required to provide in a captured scenario was name, rank, and serial number. However, the Germans tried to elicit much more information, such as group and squadron numbers, name of commanding officer, type of aircraft flown, bomb load, defensive armament, how they were shot down, names of crewmembers, target the day of being shot down, and mission. In turn, the German government would relay the information to the Red Cross, the Red Cross would send it to the United States government, and they would notify the prisoner's family.

If a form was not filled out, the German officer would sometimes accuse the prisoner of being a spy and emphasize that spies in Germany were shot. Signing the form ensured his family would be notified of his capture. Some individual interrogators used threatening language predicting solitary confinement on starvation rations unless the POW would talk. Despite the pressure and distress, prisoners refused to sign the forms. So, the officer would leave the cell, but he sometimes returned to try again.

The Dulag Luft interrogators spoke perfect English, as many had lived in the United States before the war, and they were professional. They developed sophisticated interrogation techniques that included repetitive questioning. The interrogators and German officers were conversational and apparently friendly. They sometimes offered cigarettes—a prized treat! Individual interrogators might have had a friendly demeanour at first, but in frustration from lack of answers to their questions, they could turn angry. They also employed food and sensory deprivation and took advantage of continued shock. Violence was not utilized as a form of torture, although there was an occasional slap on the face dealt in the heat of anger.

It is unimaginable what effect this whole process must have had on a young pilot, who was miles away from home, having already suffered the shock and fear of a crashed aircraft, exhaustion, hunger pains, and, in many cases, severe nicotine withdrawal. Obviously, he would want his parents or his wife to know that he was still alive. "What harm could that do?" he might rationalize. However, such cooperation was the start of a slippery slope that could lead to more information being supplied or lies being detected. It was an incredibly difficult situation that new prisoners found themselves in.

The men were isolated in small cells and taken out several times to meet with an interrogator who wore a Luftwaffe uniform. Their meals were meager; rations were two slices of black bread and jam with ersatz (imitation) coffee or tea in the morning, watery soup at midday, and two slices of German brown bread made with sawdust at night. No International Red Cross parcels were issued, and if they were, they were generally withheld to coerce prisoners into providing information.

Further interrogations often revealed that the Germans already knew a great deal about squadrons, crew, strength, aircraft type, station, and station commander's name. They probably just lacked bomb load and fuel load. That would have shocked the new POWs and probably made them wonder whether or not it would matter if they "confirmed or denied" information.

Within Dulag Luft's buildings, the men were surprised to discover that the interrogators knew so much about them personally. Furthermore, U.S. newspaper articles provided personal details about airmen's wives, parents, and siblings and their addresses. It is no wonder that the Germans were able to find this information because Dulag Luft had a huge database of information from a newspaper clipping service. Can you imagine a clever interrogator presenting it to a scared and tired prisoner? The questioning could go something like: "We know that your wife is Dolly Smotherman and that she is living at 2106 West Tenth Street in Dallas with her parents. Don't you want to protect her?" This would be terrifying news to the prisoner and a serious blow to his morale and strength.

It was a disturbing time, and Richard "Dick" Winn, for one, alternated between despondency and near panic and hope, as he stated, "These mood swings were very frequent." The limited rations given to the airmen compounded their distress.

In his diary, George Archer recorded the three days he spent in the Dulag Luft cells. He briefly noted the conditions of his accommodation and treatment. "In cooler. Bloody awful. Nothing to do but sleep." The next day, he was "still in cooler, very hot. No windows open. No books or shaving. Sleeping is the only saviour. Five dirty spuds and soup for dinner. Bread and herb tea."

Fortunately, the prisoners had been trained on what to expect, including the bogus Red Cross officer who gave them the deceptive questionnaire to fill out with the threat that their parents would not be told if they were dead or alive unless the form was fully completed. Most POWs withstood the harsh treatment and revealed no important military information other than name, rank, and serial number. Despite the conditions, the POWs fell back on their sense of loyalty and patriotism to weather the extreme conditions.

The duration of a POW's solitary confinement at Dulag Luft depended on whether he gave any information or not. Those who refused to give anything were processed much quicker. Those who showed fear or who were high-ranking pilots, for instance, were held longer. The latter were often wined and dined and taken on excursions.

Ben was interrogated there for three days at the beginning of August 1943. His copilot, Dave Pollak, was there for eight terrible days. The Geneva Conventions specified a maximum of thirty days for interrogation, though few prisoners stayed that long.

After a POW showed he was of no use for information and the interrogation was completed, he would usually be sent to the main compound of regular barracks staffed by permanent POWs selected by the Germans. There, he was allowed to go outside in the barbed wire enclosure, visit with friends, and receive regular meals made from Red Cross parcels delivered there. The Germans held the men in the barracks, fed them well, and gave them clean clothes through the Red Cross. When they had a sufficient number to ship out, they typically did so on trains.

After being in the main compound, my father and the other prisoners were assembled at the transit camp in the Oberursel location to be transported by train to their permanent camp. They usually travelled as a group, normally packed in prison cars attached to a passenger train. Windows were covered so they could not see the bomb damage in all the cities they passed through. Again, conditions were abhorrent and disgusting with little or no food, one dirty toilet per car, and armed guards keeping a close eye and locking each car. Later in the war, cars were so crowded that some prisoners had to lie down in the luggage racks above the seats. Some stood, and some sat down on hard benches. Then, they would rotate positions to share the misery.

My father's journey across Germany in that train took two days. Finally, he arrived at Sagan in "Niederschlesien" (Lower Silesia) in Germany near the Polish border, approximately one hundred miles southeast of Berlin. The railway station at Sagan was a quarter mile from the prison camp. Weary POWs walked the distance down a rutted dirt road to the camp, and some were bused. To their right along the way were tall, stately pine trees as far as the eye could see. On August 5, 1943, my father arrived at his new home, Stalag Luft III.

Chapter Seven
Behind the Wire

As Ben Smotherman arrived at the main gate of Stalag Luft III on August 5, he was tired, scared, and apprehensive of what his new home would be like. The first view he saw was overwhelming: a stark camp, barbed wire, guard towers, dirt ground, and weathered gray barracks. Likely he had not seen his fellow crewmates since they had gone down in Holland. He did not know where they were, how they were being treated, or what their future held. Officers were imprisoned at different prison camps than enlisted men or non-commissioned officers, so the likelihood of seeing the six non-commissioned officers on the crew was slim to none. Only the pilot, copilot, navigator, and bombardier from the B-17 aircraft and similar officers from other aircraft would be sent to the compound in Stalag Luft III where my father was, or to Stalag Luft I.

A pilot or commander was not just the main flier of the airplane or ship; he was also the leader of the crew, inspiring unity, care, and teamwork in a group of ten heretofore disconnected men. So, I am sure Daddy Ben was very concerned about the rest of his crew and hoped and prayed that they were all okay.

Journalist Andy Rooney was a writer for the *Stars and Stripes* Army newspaper and covered the 8th Air Force from his office in England. He even flew as a gunner on a B-17 combat mission! As he wrote in his book, *My War*, "Pilots were heroic out of proportion to their numbers. They were almost always the most capable, most intelligent, and most determined to accomplish the mission of anyone on board. They were in charge for good reason. Unlike so much authority in the military that is based on nothing but rank however attained, the pilot was in charge because he was the natural leader on board... There were a lot of pilots and crewmembers of the 8th Air Force who did very brave things. Bravery is closely associated with death because, without the possibility of it, bravery is less extreme."

Through the trying times of flying in combat, being shot down, trying to evade capture, being arrested and thrown into a German prison, enduring solitary confinement and interrogation at Dulag Luft, and now being thrust into a new home with an uncertain future, I believe my father was more than a pilot. He was also a courageous leader.

German prison camps in WWII were organized by the prisoner's branch of service, so the Stalag Luft (abbreviated from *Stammlager der Luftwaffe*) prison camps were operated by the German Air Force. The Luftwaffe treated its captives well but not nearly as well as the Americans treated all the German POWs who were sent to every state in the United States. The Wehrmacht and the Kriegsmarine (German Navy) had their own POW camps.

Stalag Luft III was carved out of a dense pine forest where trees had been planted in even rows (Figure 7-1). The first compound, East Compound, opened on March 21, 1942, and the first POWs were RAF and Commonwealth officers who arrived in April 1942 from various prison camps. There were also some American prisoners who flew with the RAF before U.S. air bases were built in England and before the U.S. entered the air war in the ETO.

Figure 7-1: A view across Stalag Luft III showing tree stumps where the pine forest once stood (photo courtesy of Clark Special Collections Branch of the USAF Academy Library)

At the time of my father's arrival in August 1943, there were three compounds: East, Centre, and North. In June 1943, all but fifty of the British non-commissioned officers living in the Centre Compound were transferred to Stalag Luft VI in Lithuania to make room for the ever-increasing number of American airmen. Early on, an enlisted man could be sent to Stalag Luft III to serve as a "batman" (servant) for an assigned British officer. But more space was needed as the war wore on, so they were transferred. USAAF personnel then changed the British spelling to Center Compound, where Ben Smotherman was initially imprisoned.

Although Stalag Luft III was commanded by German officers, each compound had a senior officer to communicate with the Germans. The highest-ranking officer in each compound became the Senior British Officer (SBO) or the Senior American Officer (SAO). Only the highest-level American and British officers were allowed to interact with the German hierarchy of the prison camp.

One notable American was Lieutenant Colonel Albert P. "Bub" Clark, whose Spitfire was shot down on July 26, 1942, on his first mission. Clark would emerge as the first American SAO from his arrival on August 15, 1942, until March 1943, when higher-ranking American officer Colonel Charles G. Goodrich entered the camp and became the SAO. As a lieutenant general, Clark later became superintendent of the United States Air Force Academy, and his extensive POW collection is now housed there.

Each compound, or "lager," was laid out in a similar way and consisted of barbed wire fences twelve feet high with another wire fence six feet inside of the outer wire fence. The fences ran parallel to each other, and concertina razor wire lay on the ground about three feet high in between them.

At each corner and at regular intervals about one hundred yards apart, there was a guard tower (Figure 7-2). The POWs nicknamed the guards "goons," a somewhat derogatory name taken from a British cartoon. So, the guard towers were known as "goon boxes." The goons also patrolled the grounds of the prison camp, conducted roll call, made surprise inspections of the barracks, and used trained guard dogs.

When my father passed through the gate of Center Compound, he was greeted by a swarm of "veteran" prisoners who were checking out the new crop. The new prisoners were shocked at the unkempt and worn look of the guys who had been in prison for a while, but after the ordeal that my father and the other prisoners in his group had been through, they probably didn't look that great either!

Figure 7-2: The guard tower or "goon box"
(photo HU21018 courtesy of IWM Duxford)

The groups of new prisoners were termed "purges." They were not yet accepted and trusted by the current prisoners because the Germans were notorious for planting "fake" prisoners to spy and gain inside information. So, before any one of the new prisoners was trusted, they were questioned and asked things that only a true American would know, such as "Who won the last World Series?" If a new prisoner could find a man who knew him and could vouch for him, he was automatically accepted. For example, when my father's friends Casey and Baker were shot down on a mission over Germany on August 12, 1943, they joined my father at Stalag Luft III South Compound, so Daddy Ben immediately vouched for them as being true red, white, and blue Americans. When new prisoners passed this screening, they were let in on Allied airmen secrets and potential escapes. If they did not pass this screening, they were followed around the camp until it was determined whether they were German "plants" or real Allied prisoners.

The new prisoners were given photo identification cards showing their barrack and room number and were also provided square metal dog tags showing their new POW number. They were fingerprinted and, after being processed, assigned to their barrack. My father was given POW number 1922 and was assigned to barrack 133, room three. In the barracks, the POWs pooled items from their own Red Cross parcels to provide a first meal. Each prisoner was supposed to receive a special American Red Cross suitcase containing a sweater, pajamas, a hat, gloves, a shaving kit, a sewing kit, socks, soap, a toothbrush and toothpaste, cigarettes, and gum.

The POWs were given complete freedom to mingle with the other POWs and were allowed to select their own roommates. Lucky ones lived in a room with several old friends whom they unexpectedly met at the camp and also with members of their crew.

The Geneva Convention of 1864 created the International Red Cross, which has the responsibility of ensuring humanitarian protection and assistance for victims of war and other situations of violence. During WWII, they provided food, clothing, medical supplies, and mail. Hand in hand with that, the YMCA was responsible for providing recreational, educational, and religious supplies. The YMCA also made routine visits to all the POW camps and took pictures of the conditions. I am forever indebted to these two agencies for their crucial role in ensuring the mental and physical well-being of my father and the other POWs as well as keeping the families informed of the status of the prisoners.

Each compound (Figure 7-3) consisted of ten to fifteen wooden single-story huts, and each ten-by-twelve-foot bunkroom initially slept four to eight men in double deck bunk beds. Every room had a window, a small freestanding stove (fueled by coal or wood) in one corner for heat and cooking, a single hanging lightbulb, a table with wooden benches, and a locker for each prisoner. As the war progressed and more POWs arrived, another layer was added to make triple deck bunk beds (Figure 7-4).

There was a central aisle the length of the barrack with an entrance door at each end. Every barrack had a washroom (with twelve to sixteen cold water faucets) and a small communal kitchen with a sink, cold water, and a coal stove. There was no main dining room, so the prisoners ate in their rooms and prepared their own food (except for German brown bread and one bowl of soup per day provided by the Germans).

Figure 7-3: Typical layout in Stalag Luft III. Note the distance between the huts and the wire (bottom left) designed to expose any escaping prisoner (photo HU21013 courtesy of IWM Duxford)

Figure 7-4: Typical prisoner room with the third bunk that was added in 1944 due to the increasing number of prisoners in Stalag Luft III (photo courtesy of Clark Special Collections Branch of the USAF Academy Library)

There were four corner rooms, three of which housed two senior Allied officers each. The fourth corner housed the non-flushing toilets for nighttime use because the prisoners were locked up inside at nightfall. According to navigator Bob Doolan, who was in Center Compound, the first person to use the indoor toilet at night had to clean it out in the morning. As he told the story, everyone was "holding it in" until the first man used the toilet at night. Then, after that, there was a long line!

During the day, prisoners used outside pit latrines (called aborts) in the compound. Some of the POWs also had gardens where they could grow vegetables or flowers, but most produced almost nothing in the yellow Silesian sand.

Stalag Luft III was specifically designed to be "escape proof" based on what the Germans had learned from having POWs for two years. For example, the barracks were raised from the ground eighteen inches so that "tunneling" could be detected. The only part of each barrack that was not visible by this eighteen-inch clearance were the washroom and the stove locations, which had brick and concrete built up from the ground to house the pipes and drains of the washhouse and to provide structure for the heavy stove. Furthermore, the barracks were several feet from the perimeter, so any tunnel would have to be very long.

Each prisoner had at least one uniform, often sent from home, and one good pair of shoes. They did their own laundry in buckets, plunging the laundry up and down with a "Dobie" stick created for such a purpose. Some used basins in the washroom. However, sheets and towels were laundered every two weeks in a town next to Sagan. Prisoners themselves were allowed one hot shower a week.

My father had a lot to learn and had a lot of questions about being a POW at Stalag Luft III. *What will the food be like? How do I combat boredom? How do I communicate with my family? What are the other prisoners like? How will I be treated by the Germans? What is the USAAF hierarchy like?*

My father was exhausted after nine days of capture, interrogation, travel, and food deprivation. Sleeping accommodations were not the best. Although the single mattress on his bunk bed was filled with lumpy straw, he probably slept hard those first few days. Being in a compound of Allied prisoners with other pilots and officers, he was full of hope for a better future than the week he had just experienced.

What a world awaited this young airman, Ben Smotherman. Captivity had to be experienced to be fully understood. He could no longer do what he was trained to do—fly airplanes and defend his country and the other Allied countries. In addition to his apprehension and exhaustion, he probably felt some guilt in surrendering to the Germans. He had little communication with his loved ones, had no freedom outside the barbed wire, lived with men he was forced to get along with, had no privacy, shared crude bathroom facilities, never had enough to eat, didn't have enough clothing or shoes, and was subject to the broad sweep of a bright spotlight from the goon box moving across the barrack windows all night long. He was stuck where he was, but the military had told him that was not heroic or commendable. In fact, it was the duty of POWs to attempt escape from the prison camp.

The "intelligence" entity of the organization utilized Allied senior officers in the camp who worked hard to facilitate the acquisition of escape materials. Clandestine cameras were created from smuggled parts from packages supposedly from families. Radios and compasses were made out of scrounged materials and those obtained by bribing the guards. Artists in the camp helped to create fake passports. money, and travel documents bearing photographs for realistic identification cards. Others could draw maps, make civilian clothes, and do whatever was necessary to assist the escapee. About 60 percent of the Kriegies were involved in some type of activity related to escape.

Once my father was one of the "trusted" prisoners, he was made aware of the escape committee and its function. With his artistic talents, I suspect that he helped create some of the items, although he was not listed in Bub Clark's notes as having helped the escape effort.

Every morning at 9:30 a.m. and every evening at 5 p.m., the prisoners were lined up in a roll call, called appell in German, held in a large, clear area called the *appellplatz*, which was also used as a sports field. Each prisoner had to be accounted for. The prisoners marched out in military fashion by block/barrack and lined up. When a particular group was being counted, they stood at attention for the German officers.

It was at the first appell that my father met his fellow prisoners, many of whom had established the rules and organization that were followed by subsequent compounds. Some of the men had been shot down in their aircraft, like my father, and some had bailed out in their parachutes. Some had escaped from burning aircraft, and some had experienced wrath and punishment by angry citizens of Germany and occupied countries. Some had landed in water. Some had narrowly escaped death and witnessed the death of others on the crew. When fliers congregate, they talk about their missions, their airplanes, their crews, and, in this case, how they ended up as a POW. Oh, the stories that they must have told and shared!

After appell, prisoners were allowed their own free time. On that first day in prison, Ben took it all in and became familiar with the duties, organization, military hierarchy, and rules laid out by the Germans.

My father wasted no time in writing to his family and, presumably, my mother. Keeping them short and sweet, he wrote on designated postcards, and when received, my aunt typed them for distribution to the family. The first one that Ben wrote was dated August 6, the day after he arrived at Stalag Luft III. It said:

Dearest folks, I guess you know where I am now, but things aren't as bad as you probably imagine. I'm feeling fine and didn't get scratched. I'm getting enough to eat, plenty of rest—living conditions are okay. Hope you all are okay. See the Red Cross about writing to me. Love, Ben

However, his family did not see this postcard until December 18, after it had been through American and German censors and through the Red Cross. The high volume of mail that went by train was slowed every time the Allies bombed the railroad tracks. Other short letters were mailed by Ben about once a week in August and September but were received about a month later. Despite these letters written by Ben in August and September, his family still did not know whether he was a POW or even whether he was alive.

By the time he arrived, the British and Americans in North Compound were working very well together. They trusted each other and worked on escape strategies together. During that time, plans for the Great Escape were born in North Compound. However, the compound was quickly filling up with prisoners, and in the summer of 1943, the Germans decided to build South Compound, which would house only Americans. North Compound would then only house the RAF and British Commonwealth airmen.

Later in August, Roland L. Sargent, who piloted the B-17 on my father's first combat mission, was shot down over Belgium after bombing a ball bearing plant at Schweinfurt, Germany. He evaded capture for ten days but was captured by the Gestapo in Paris and then put in the notorious Fresnes Prison as a spy for one month. After that, he arrived at Stalag Luft III.

I can picture what a reunion Sargent and my father had! Roland Sargent signed my father's Wartime Log as one of the prisoners in Stalag Luft III on the 92nd BG page. They were undoubtedly glad to meet again.

There were many stories to tell, and there was much catching up to do. Sargent filled my father in on all the recent activities of the 92nd BG and the 327th BS and the status of their mutual friends in their shared squadron. The 327th BS had moved from Alconbury to Podington in September 1943, unbeknownst to my father. If he had not already found out, Sargent probably told my father that the last mission of the YB-40 was the day after my father was shot down. The YB-40 turned out to not be effective as a bomber escort, and the extra armament and lack of bombing capability restricted its usefulness.

Sargent probably also told my father about the upgrade of the B-17F to the B-17G, which entered service in the fall of 1943. To help with the head-on attacks by the Luftwaffe, the B-17G added a chin turret, a design that was in the YB-40, below the bombardier station in the front of the aircraft. It was a successful design change made to the B-17 as a result of the "test flights" of the YB-40 that my father made. The staggered waist gunner positions that were a part of the YB-40 design were also incorporated into the B-17G. Another welcome addition was the inclusion of glazed windows in the waist gunner positions, which helped to alleviate the dangerous conditions of below freezing temperatures and gusty winds at flying altitudes.

The newly imprisoned airmen who arrived at Stalag Luft III let Ben know that the armament of the German fighters had reached maximum capacity. The cannons were heavier and fired more rapidly, and the underwing mountings for the rocket launchers added to the deadly capability of the Germans.

The new South Compound opened on September 8, 1943. All but one of the American prisoners from North Compound, most of the American officers from Center Compound (including my father), and the American prisoners from the German POW camps in Italy were transferred there. As a result, when South Compound opened, it housed about 1,200 American officers. Although the senior officers from North and South compounds communicated with each other, the POWs between the compounds no longer had direct communication.

Colonel Goodrich, the SAO of South Compound, entrusted the "security and intelligence" function to Bub Clark. The security and intelligence officer was in charge of any escape plans, and any ideas for escape had to be approved by him and his committee. Meanwhile, as most of the American POWs were consolidated into one compound, Colonel Goodrich, known as "Rojo," had a seasoned group of prisoners and set about organizing South Compound.

South Compound was organized by Rojo and his senior officers so that administrative functions ran smoothly and efficiently. Examples of administrative functions included kitchen duty and jobs ranging from mail officer to laundry officer. Some POWs were responsible for organized sports and entertainment. According to Malcolm Higgins, a B-17 pilot imprisoned at Stalag Luft III, the worst parts of being a POW were boredom, inactivity, and confinement. So, once social and sports activities were organized, morale rose throughout the entire camp. The jobs and activities kept the compound organized and also kept the prisoners busy. It was key that everyone was coordinated and scheduled due to the close quarters and eventual overcrowding. The commandant of the camp, Colonel Friedrich von Lindeiner, approved of the activities, believing busy prisoners would not think about escaping.

Rojo kept strict military discipline in South Compound, implementing a military-style marching and formation at the daily appells. My father was pictured in a photo taken by the Germans of the appell in South Compound (Figure 7-5).

Figure 7-5: Roll call (appell) in South Compound (German photo courtesy of Clark Special Collections Branch of the USAF Academy Library)

As South Compound and its prisoners settled into the "new normal," it was important for the POWs to have religious services. A much-beloved chaplain, "Padre" Murdo MacDonald, who conducted church services on Sundays and performed clerical duties, was known to all compounds. Interestingly, he was a Scottish paratrooper who was shot down and captured in North Africa.

On September 5, 1943, my mother again received an official Western Union telegram hand delivered by a messenger. It stated: "REPORT RECEIVED THROUGH THE INTERNATIONAL RED CROSS STATES THAT YOUR HUSBAND FIRST LIEUTENANT BENJAMIN F SMOTHERMAN IS A PRISONER OF WAR OF THE GERMAN GOVERNMENT LETTER OF INFORMATION FOLLOWS FROM PROVOST MARSHAL GENERAL. ULIO THE ADJUTANT GENERAL." Now, my mother had official word that my father was a POW in Germany. She'd had over a month of uncertainty as to whether my father was alive or dead, on the run with the help of the underground, or captured as a prisoner.

At about the same time my mom received this telegram, my father was settling into his new barrack and getting to know his roommates right after South Compound opened. He was a veteran Kriegie of one month and had become accustomed to the food and lifestyle in his new quarters.

His family received a letter written September 26, but it did not arrive until December 31:

I've moved to a new camp, and it's much better here. I'm still feeling fine and hope all of you are the same... I'm still reading a lot and going to school some. Time passes fairly fast. Hope to be back with you before too long. Remember I'm thinking of all of you. Take it easy, and don't worry!

In Ben's letters home, he always remained courageous and encouraged people to not be "down in the dumps." All in all, he stated that his morale was good. In his letters home, he also made a point that he had been busy drawing pictures for fellows and had received cigarettes in payment! Of course, he thought that the war would be over soon and that he would be coming home. He kept meticulous notes about things that he had drawn or painted, including signs that were put up around the prison camp.

As in the other compounds, South Compound had athletic fields for sports activities. The POWs had been kept busy digging up tree stumps and roots to create the fields. Available sports included volleyball, base-ball, softball, touch football, badminton, deck tennis, basketball, and boxing with leagues organized for most (Figure 7-6). Furthermore, POWs played ping pong, wrestled, lifted weights, and exercised on horizontal and parallel bars. Sports equipment was provided by the War Prisoners Aid of the Young Men's Christian Association (YMCA). In addition, pris-oners sometimes swam in the fire pool, a structure in the ground at each compound with a few feet of water to use in case there was a fire in the camp. Another common exercise was to "walk the circuit," which was the safe perimeter of the compound (Figure 7-7).

Figure 7-6: POW boxing tournament (photo courtesy of Clark Special Collections Branch of the USAF Academy Library)

Figure 7-7: Prisoners walking "the circuit" around the perimeter of a compound. Note the thin "warning wire" about thigh high (photo courtesy of Clark Special Collections Branch of the USAF Academy Library)

My father wrote in his POW Wartime Log:

The sports we had in Stalag Luft III did more than satisfy our desire for competition. They took our mind off the war and helped to pass time quickly.

Softball held the spotlight in camp throughout the summer, with basketball, volleyball, and deck tennis secondary. We had two softball leagues, majors and minors, and each block had a team in each league. Competition was keen, and the following large. Good sportsmanship prevailed. We were sorry to see the softball season end, for we realized how it helped to pass the time, and winter was near at hand.

Without a doubt, Ben was eager to paint and draw, and he drafted an excellent rendering of South Compound, which showed the amenities and facilities that existed at the time (Figure 7-8). The diagram also showed the washhouse, the pit latrines, the cookhouse (used by the Germans to cook the thin, watery soup they provided), various athletic facilities, and the fire pool. Also depicted was West Compound which was added in July 1944 for American officers. A red arrow points to Ben's barrack 133 in the middle of the compound.

Figure 7-8: Ben Smotherman drawing of South Compound. Arrow denotes his room in barrack 133

The drawing also shows the shower house. The men used tin cans with holes punched in them and ran water hoses or faucets into them to create some semblance of a shower! When I met Bob Doolan, who remained in Center Compound, he said they did not have hot water, but South Compound did! The 105-year-old hero told me that more than once, so I guess the Center Compound POWs were a little bitter about not having hot showers! However, the showers in South Compound were not operational until August 1944.

My father drew a picture of his overcrowded room (Figure 7-9), which, at the time, housed twelve prisoners. A chart drawn by my father in his logbook lists the names of his roommates and their addresses (Figure 7-10).

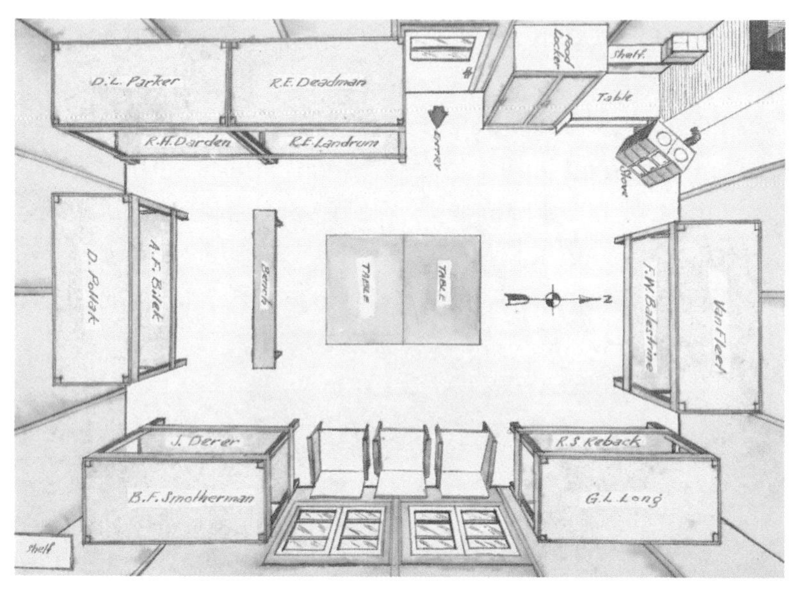

Figure 7-9: Ben Smotherman drawing of his room in barrack 133

As South Compound became organized, crewmate Dave Pollak took over the job of parcel distribution for South Compound, sorting Red Cross parcels, boxes from families, and any supplies that were delivered. As a result, Pollak was able to detect packages with a special marking that denoted escape paraphernalia was inside. For example, they received a specially marked parcel of baseballs manufactured by the P. Goldsmith Sons Company from Pollak's hometown of Cincinnati. The baseball's interior had a metal sphere that contained camera or radio parts instead of the typical cork center.

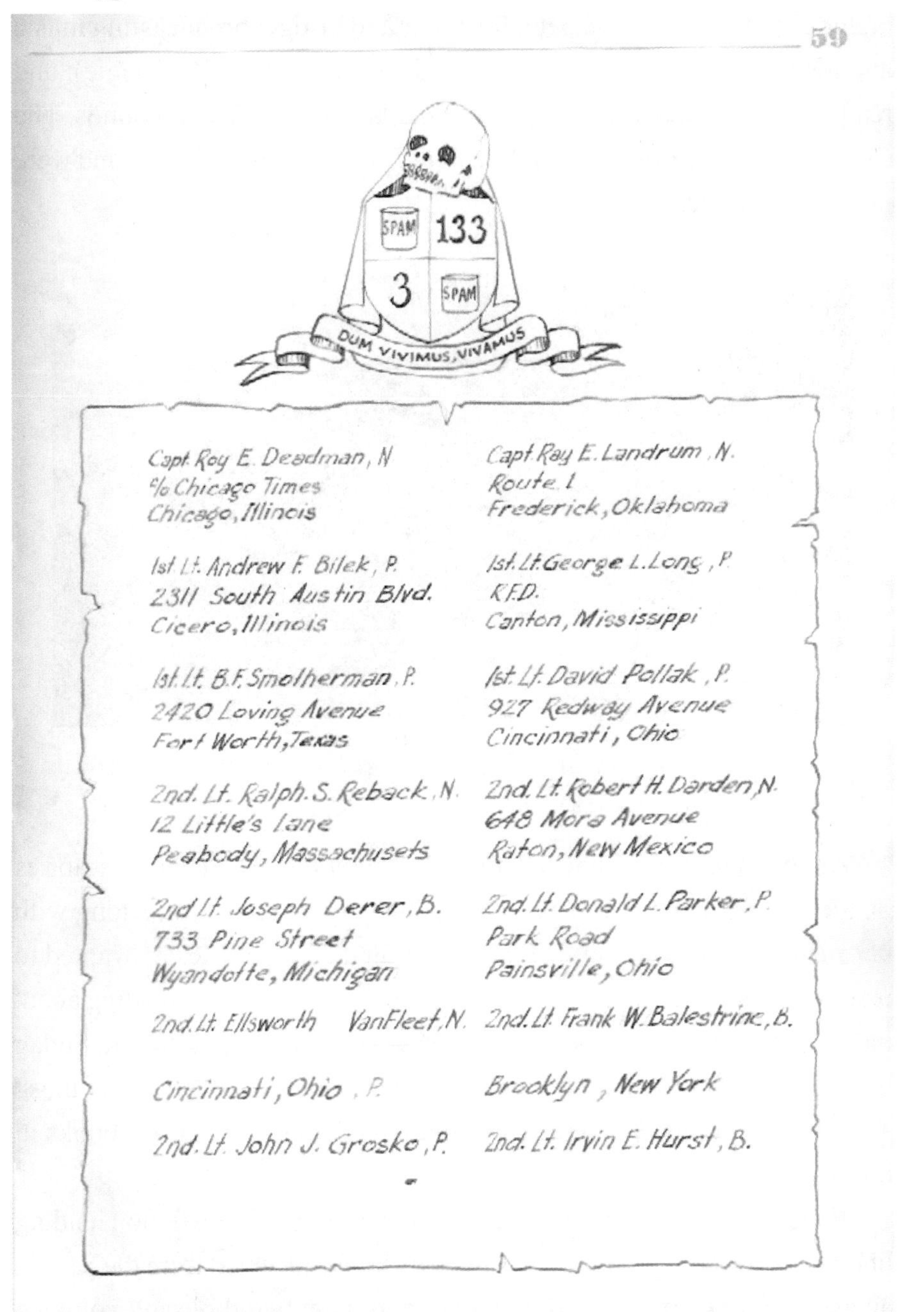

Figure 7-10: Ben Smotherman drawing of his roommates in barrack 133, drawn in his Wartime Log

Prisoners at Stalag Luft III also had a very well-stocked library (Figure 7-11) and other activities to combat boredom and maintain morale. Those included playing cards, especially poker and bridge; broadcasting music and news over a camp amplifier called "Station KRGY" (Kriegie); and reading newspapers created by individuals in all POW compounds. The *Circuit* in South Compound and *Kriegie Times* in Center Compound were just two examples.

Figure 7-11: Library in Stalag Luft III (photo courtesy of Clark Special Collections Branch of the USAF Academy Library)

Although the prison had a library, my father requested books—stories on the sea, art, history, or any topics—because he read a lot. Along with books about aviation, most of the books on his list of titles he wanted to read were about mariners or the sea. In addition, he enjoyed reading about the American West, cowboys, and Texas. Thanks to the library in Stalag Luft III, there were many books that he read, and he kept a list of those as well. Sometimes, he kept interesting facts or notes from the books in his notebook.

Reading was the greatest single activity of POWs. The fiction lending library of each compound was enlarged by books received from the YMCA and sent by next of kin until it totaled more than two thousand volumes. Similarly, the compounds' reference libraries grew to include over five hundred works of a technical nature.

In addition, some prisoners conducted classes for the other prisoners, and degree programs were available from the London University or the Armed Forces Institute of the U.S. Army. Popular options included French, Spanish, and German language classes as well as accounting, navigation, banking, law, theology, and engineering. The Education Department's classes ranged from aeronautics to law, painting, sketching, economics, mathematics, history, and other subjects with prisoners serving as instructors. Many POWs earned degrees, and the exams were supplied by the Red Cross and supervised by academics such as a master from King's College who was a POW there. Ben started taking classes and was planning to teach art, drafting, and aerodynamics to the other prisoners.

Foremost in the POWs' minds was food and the preparation of their meals. As evidenced by much discussion in various Wartime Logs, the prisoners focused on food more than anything else. According to an interview with Bub Clark, sex and women were prime subjects of conversation only when food was abundant. He could tell when the prisoners were hungry by how much they talked about food instead of women.

The German ration of food for the prisoners was inadequate. Food provided by the Germans consisted of black bread, lots of cabbage, turnips, fibrous inedible kohlrabi, ersatz margarine, foul-smelling fish, cheese, barley, runny ersatz jam, rotten potatoes, blood sausage (no more than congealed blood with a few slices of onion in a casing), and ersatz coffee made from acorn shells and other mysterious ingredients.

According to Ben van Drogenbroek's excellent book, *The Camera Became My Passport Home*, the highly secret German recipe for black bread was 50 percent bruised rye grain, 20 percent sliced sugar beets, 20 percent tree flour (saw dust), and 10 percent minced leaves and straw. My father's notes indicated that German bread was used to make bread puddings and was burned as fuel! Men toasted it on the side of the hot stove in the barracks.

Breakfast was usually tea or coffee and the black bread toasted with margarine and jam. For lunch, four days a week, German soup was prepared in huge vats in the cookhouse by the Germans, and POWs from the barracks were sent to pick it up. Inferior products, such as rotting potatoes, weevil-infested grain, or animal carcasses, were often used. Maggots and worms found their way into the soup for "protein." Meat was scarce unless a horse died. It is interesting to note that their rations were the same as those issued to German soldiers.

To supplement the German rations, each prisoner received a weekly box from the International Red Cross, which contained an assortment of food, such as nonperishable biscuits, raisins, coffee, chocolate, powdered milk (KLIM—milk spelled backwards), canned beef, SPAM, and fish as well as other necessities like cigarettes and soap. The content of a typical box is shown in Figure 7-12.

Figure 7-12: Contents of typical International Red Cross parcel (courtesy of USAF Museum)

During the war, over twenty-seven million Red Cross parcels were delivered to the American and Allied POWs. The parcels were shipped from Geneva, Switzerland. In South Compound, the SAO required that all contents of the Red Cross parcels be pooled so that a fair and even distribution could be made among the prisoners. Chocolate and cigarettes were the exception. Each man received one parcel per week with one half received on Wednesday and one half on Saturday. However, as the war progressed, logistics and problems with transportation resulted in prisoners only receiving on average one half of a parcel a week or less.

The most popular commodities from the Red Cross were cigarettes and chocolate. They were used as currency in the camp, as shown in Figure 7-13, and traded in a bargaining process. Chocolate D bars were rewards to those who won card games or league teams who won championships. My father saved labels from three different types of chocolate bars (English, Swiss, and Argentinian). Now I know where I inherited my love of chocolate!

Figure 7-13: POW card game where chocolate and cigarettes were used to make bets (photo courtesy of Clark Special Collections Branch of the USAF Academy Library)

In addition, medical supplies, medicines, glasses, and dental supplies were provided by the Germans but supplemented by the International Red Cross when the need arose. The infirmary was therefore well equipped, and regular medical care and typhoid shots were available. A medical team of doctors, both POW and German, took good care of the Kriegies, but dental care was marginal with many POWs getting teeth pulled by a fellow POW.

After a month in South Compound, Ben and the other Americans liked the fact that they could organize themselves and be among the other USAAF airmen in a U.S. only compound. He was setting up in his new home, cementing his relationships with his roommates and fellow prisoners, and finding ways to keep himself busy. He preferred to smoke a pipe rather than cigarettes, so he carved a pipe for himself in his beginning days in the prison camp. In early correspondence with his parents and sister, he wrote that he had been making sketches of his fellow prisoners and receiving cigarettes for pay and that he was working on a small wooden model of an old-time stagecoach. By November 1943, things were looking up, as routines and food distribution had been established, communication from friends and family had begun, and information about the war was being quietly disseminated.

As an example of good things that became available, the POW Wartime Log journals were provided by the YMCA and distributed by the International Red Cross. They were blank books with 151 numbered pages and twenty un-numbered black scrapbook pages in the center section for photographs or newspaper articles.

To his delight, my father received a Wartime Log, along with colored pencils, watercolors, and graphite pencils, on November 16, 1943. I vividly remember when Daddy Ben first showed his Wartime Log to me. That was the only time he ever told me about the war, being shot down, or being a prisoner. I was in awe of his drawing, writing, and painting ability, and I knew that the book was a treasure and a piece of history. In it, he included not only paintings, drawings, and stories but also autographs and addresses of fellow prisoners and musical scores of songs. He painted different types of WWII aircraft (from memory), and aircrew prisoners autographed the pages of their associated aircraft. Of course, not everything that went on at the prison camp was described in the logbooks, as they were frequently censored and checked by the Germans. Nothing negative could be said about the Third Reich. But Ben's logbook did offer a unique view into his life as a POW.

My father's family and my mother started receiving Red Cross POW bulletins, which gave a good accounting of what was going on at the various prison camps around the world during WWII. The Red Cross duplicated newspapers that were created by the POWs and included them in the bulletins. My mother was given an address where she could correspond with my father and directions on how to send packages. Unfortunately, the only correspondence that I have is the very first letter my mother sent to my father in prison camp. All of her other letters, as well as my father's, have been lost or disposed of over the years.

The first letter that my mother wrote was dated Monday, September 13, 1943. She had just received his address—more than two months after Ben had been shot down. In the letter, she told my father how she had tried to be brave during the month that he was MIA when she didn't know whether he had survived. She very lovingly told him, "I'm alright and have been, darling. I've tried to do like I thought you would have wanted me to. When they gave me the news, the first thing I said was, 'He's alright—he can take care of himself.' So you see, baby, I didn't let it get me down. I kept faith, honey."

Of course, this time was very stressful for my mother, and in this first letter to Ben in prison camp, she stated that she'd quit her job. After that, she went back to college to further her education. School was always my mom's happy place. I remember her later telling me that she knew Ben was a good pilot, and during the time of his MIA status, she felt in her heart that he would figure something out and be okay. She was correct in this assessment, and she knew that he was mentally and physically strong, in addition to being a multi-talented and intelligent officer.

I am so glad that my mom was staying with her parents in Dallas and that she had a lot of friends in her support network. In talking about that support to her husband, she relayed, "My darling, I didn't know we had so many friends until we needed them. Gobs of letters, telegrams, and phone calls came through. I want you to know, angel, everyone still remembers what a wonderful person you are. We're all mighty proud of you."

Like all letters the prisoners received, the letter from my mother bore the stamp of the German censor: "Gepruft 82." It was postmarked September 14, and my father received it on November 25, 1943, which was Thanksgiving Day. In his Wartime Log (Figure 7-14), Daddy Ben drew a picture of himself reading this letter with the title, "Red Letter Day"! It had been four months since he'd been shot down, and receiving a letter from his wife must have been fantastic for his morale. He wrote on that day:

Dearest folks, Received first letter from Dolly today and was so happy over it! Should be hearing from you all soon. I know it must have been a terrible month of anxiety that you didn't know about me. Hope it wasn't too bad. I'm still feeling fine. I keep warm and get enough to eat! Hope all of you are okay. Don't worry about me. I'm keeping busy! Take it easy, and write when you can!

He needed that boost in his spirits, as it was his first holiday in captivity, and he was probably feeling exceptionally homesick and missing his family gathering for Thanksgiving. Being able to communicate with one another and validate they were well under the circumstances had to be a big relief all around. In letters to his parents, Ben comforted them by saying that he was "happy and healthy." Furthermore, he wrote, "I'm not in the dumps, and I don't want you to be."

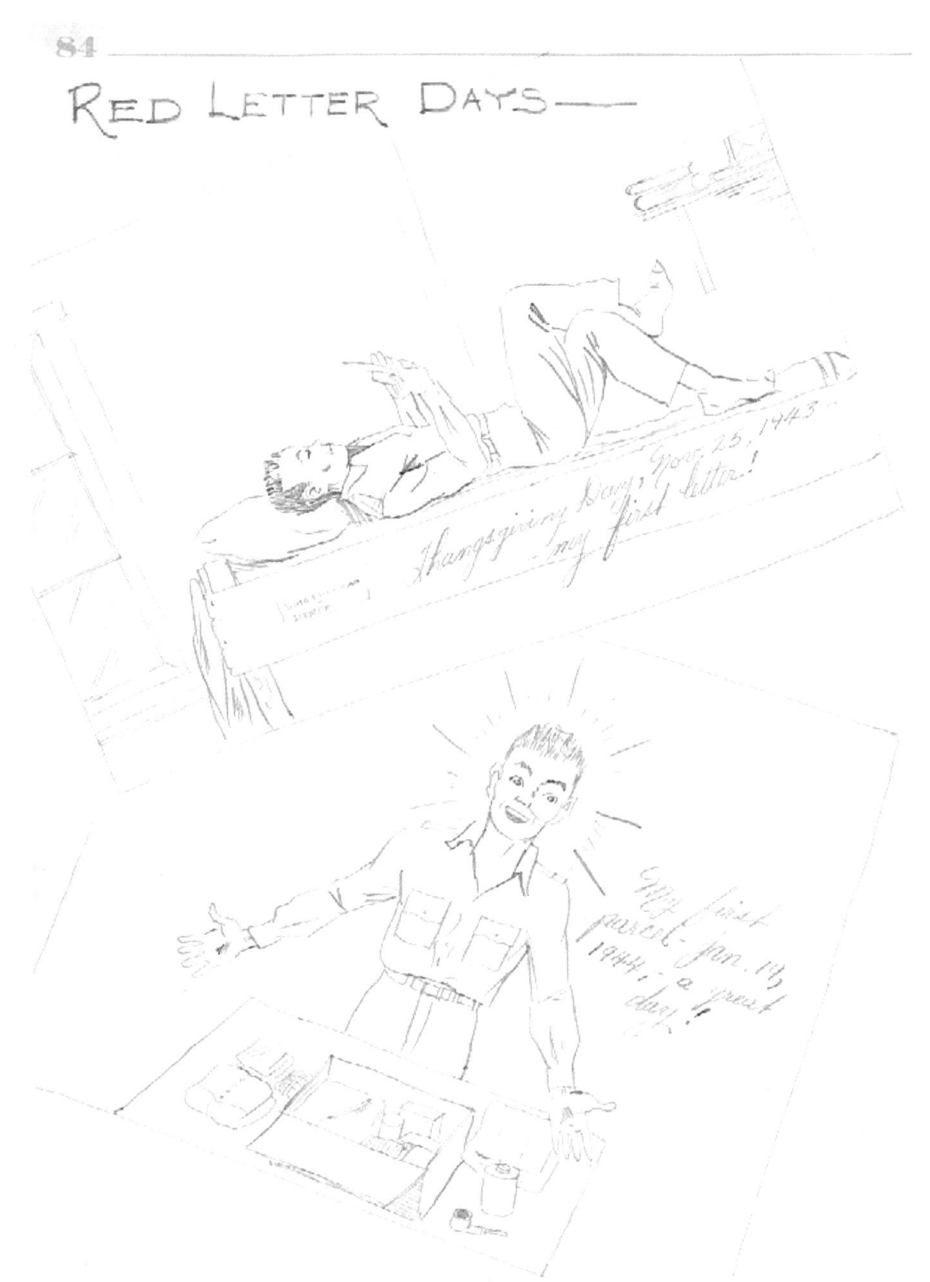

Figure 7-14: Red-letter days, as depicted by Ben Smotherman in his Wartime Log

As one can imagine, mail was very important to POWs for their morale, emotional well-being, and connection to the outside world. American POWs could send five letters and four postcards per month and could receive unlimited mail. Unfortunately, it took three to four months for a letter from the United States to reach a prisoner in Europe. Packages were also permitted, and parcels were allowed to be sent by next of kin every sixty days. The types of items allowed to be sent as well as the acceptable type of packaging were also specified to speed up the amount of time it took the Germans to inspect the parcels.

The International Red Cross made it very clear to my mother what type of stationery (onion skin, not torn), content, and lettering would be allowed. Only one page with words, preferably typewritten, were allowed on front and back. There was also a long list of subjects to avoid with the recommendation that only news of home, family, friends, and neighborhood activities be included. The Red Cross POW bulletins to the families also suggested that no more than one or two letters per week be sent so that the already taxed delivery system would not be bogged down.

The POWs had restrictions on content and could only write in block letters. Correspondence from POWs was censored by the American senior officials in the prison camp (so that no secrets were inadvertently told) and by the German censors. On the bright side, postage for all mailings and parcels for POWs was free.

Christmas was a poignant time for the prisoners, and on Christmas Eve 1943, they were not locked up in their barracks until midnight. In a letter to his wife, Stalag Luft III SAO Delmar T. Spivey said: "I hope everyone in the Christian world had as happy a time as I have. Four of your letters came yesterday! What a Christmas they [prisoners] had! Song service at 7:30 [Christmas Eve] with a Christmas story by the chaplain. Then, I inspected each barracks to see the decorations the Germans gave us and those the boys made themselves. They were extraordinarily clever and colorful. The band played in several barracks, and we sang carols and thought of home. Today [Christmas], we exchanged our little gifts; then, by permission, the senior officers visited the senior officers of other compounds. We went to church, then back for our Christmas Red Cross parcels, which were truly wonderful. We are grateful. The German officer in charge has just finished inspecting all our decorations, and now it is time for bed. Through all this day, I have been with you and Pete in prayer and thought. May God will that you both are happy and that the world be at peace before long."

The Red Cross weekly parcels to the POWs had special contents and decorative labels to celebrate Christmas. Items like turkey, Christmas pudding, ham, dates, sweets, and honey were included in a special box. Some accounts indicate that, using raisins that were included in the parcels, the prisoners were able to make raisin wine in crudely made distilleries for the holidays. They did what they could to celebrate Christmas away from home in a prison camp.

Once my father started receiving letters from home, they were a real morale booster for him and were just what he needed to make Christmas less lonely. In a card dated December 5, 1943, and received by his family on February 28, 1944, my father expressed his appreciation:

Dearest folks, I'm receiving mail from you now, and it makes so much difference! ...I'm so glad you all are okay. Dolly told me about the pipe Bill [brother-in-law] *got for me, and I'm anxious to get it. Thanks a lot, Bill—I can really enjoy it here! I'm still fine, taking care of my health, and keeping myself busy. Don't worry about me, folks. I'll be coming home before we realize it. Things could be worse, and I'm thankful.*

Allied POWs were convinced that we would win the war—it was just a matter of time. They were able to follow the war's progress via secret radios and by charting it on the map. After D-Day, they were convinced the Allies would win, and the German guards in the camp recognized that as well. Everyone was ready for it to be over. Of course, they were hoping for a quicker end to the war than what eventually happened.

With the ensuing letters between my mother and father, he was able to communicate to her what he needed. She used customs tags and forms that she completed to send a package to him. The contents listed are very telling, as they describe the kinds of things that Ben desired, what was important to him, and what the International Red Cross parcels did not adequately include.

The first customs declaration that my mother filled out was dated November 27, 1943. By that time, she had received letters from my father detailing what he would like. It really made her feel good that she was ablc to do something to help him and to make his life easier in the prison. Her list was quite long:

Pipe
Package of pipe cleaners
Bar bath soap
Terry towel
Huck (lint-free) towel
Washcloths
Box tooth powder
Toothbrushes
Plastic combs
Safety razor
Packages of razor blades
Shaving brush
Shaving mirror
Cake of shaving soap
Stick of camphor ice
Set of dice and chips
Pairs of socks
Pairs of undershorts
Undershirt
Handkerchiefs
Shoelaces
Packages of chewing gum
Sewing kit
Box of vitamin tablets
Package of safety pins
Dried soup
Bouillon cubes
Pecans
Candy bars
Cocoa
Dried beans
Sugar
Pepper
Salt
Can of nutmeg

The weight of the box was seven pounds. Ben received the package on January 19, 1944, and it was another red-letter day, as portrayed in his Wartime Log (Figure 7-14). In the drawing of the parcel, he depicted the very items listed, and in it, he had a beaming smile on his face!

In a card to his family dated January 23, 1944 (and received March 20, 1944), he wrote:

Dearest folks, Received my first parcel, and it was so nice! I'm really proud of the pipe, Bill, and really needed it! Thanks so much for it! I'm still feeling fine and takin' it easy and hope all of you are doing the same. I'm not attending or teaching any classes now but keep busy with my artwork! I'm still running a cartoon strip in the camp paper. I hope to see you before too much longer, so keep your chins up!

As the prisoners finished the 1943 year in prison after a relatively good holiday, they were convinced that they would not spend the next Christmas in prison. The new year of 1944 brought hope for better things and freedom to the prisoners' daily thoughts. However, winter in Sagan, Germany, had set in, and in 1944, it was a bitter one. Blankets were thin, and heat was almost non-existent in the barracks. The Germans supplied two blankets per prisoner, and the Red Cross supplied another blanket. Prisoners used shredded copies of a German propaganda newspaper and German magazines to insulate the blankets. They sewed them between the two German-issued blankets to make them warmer. Most of the POWs slept in their clothes because the little stoves in their rooms did not put out much heat, and the walls of the barracks were thin with no insulation. Strong winds went right through them.

Figure 7-15 gives a glimpse of how gray and solemn Stalag Luft III looked in winter. Outdoor sports were replaced by more indoor activities, such as plays, musicals, card games, reading, and studying. Ben spent most of his time drawing and painting. A few brave men created an ice hockey rink, and games were played between rival compounds.

Figure 7-15: Stalag Luft III in winter 1944; photo taken with prisoner clandestine camera (photo courtesy of Clark Special Collections Branch of the USAF Academy Library)

The theater in South Compound was completed in mid-February 1944 and held five hundred Kriegies. It had no heat but did have light, a projection booth, and an orchestra pit. The seats in the theater were made from the wooden crates in which the Red Cross parcels came. My father helped build the theater with his woodworking skills. The prisoners put on high-quality biweekly performances featuring many of the current West End shows. The theater was a popular gathering place, and performances, including theater productions, concerts, and the playing of records from back home, were scheduled every night. The YMCA even furnished a phonograph that was shared by many barracks and several records ranging from boogie-woogie to classical.

In his letters to his family, my father mentioned some of the plays he attended. He saved the program from February 18, 1944, *Strictly from Hunger*, a three-act musical complete with an orchestra and a chorus!

In addition to plays in the theater (Figure 7-16), another popular activity was concerts put on by the POW Luft Bandsters, shown in Figure 7-17. Musical instruments were provided by the YMCA. Singers could participate in glee clubs and choirs in each compound. There was even a "barber shop quartet" with four men who grew handlebar mustaches in the style of the 1890s. The performance of *The Messiah* at Christmas was one of the many remembered events by POWs, and German guards who listened outside the theatre windows had tears in their eyes. Many professional musicians from the big band era flew in the war and, when shot down, provided the same quality music as their swing bands of the time.

In early 1944, when the B-17s and other bombers had the benefit of P-51 Mustang escorts, they had the capability to bomb Berlin and other critical locations inside Germany. The POWs could see them fly overhead, which increased morale when they could cheer them on from the ground. Due to the proximity of Sagan to Berlin, the prisoners knew the bombers were able to penetrate Germany. The sight of the Allied bombers reassured the prisoners that the Allies were winning the air war. The United States could produce aircraft at a faster pace than the Germans, and the USAAF could crank out qualified pilots better than the Luftwaffe could. Therefore, the number of combat missions that an aircrew had to complete was increased to thirty and then to thirty-five before they could go home. It seemed that the Americans were gaining strength in the air war and winning the battle of the skies.

Figure 7-16: Production of For the Love of Mike play in Stalag Luft III theater, March 1943 (photo courtesy of The National Ex-Prisoner of War Association)

Figure 7-17: Luftwaffe band made up of prisoners using instruments supplied by the YMCA (photo courtesy of Clark Special Collections Branch of the USAF Academy Library)

Chapter Eight
After the Great Escape

I don't know how much my father knew of escape plans that were going on among the compounds. Throughout the war, a variety of creative escapes were attempted in the camp, from men trying to cut the barbed wire or crawl under it, to men hiding in supply wagons leaving the camp, to prisoners impersonating guards and walking out the gate. Most of these escape attempts were unsuccessful and resulted in the POW spending time in the "cooler." Per the Geneva Conventions, prisoners could not be punished for trying to escape.

In early 1944, there were elaborate plans to escape from North Compound by digging three tunnels named Tom, Dick, and Harry. By then, North Compound was inhabited by British POWs, but when there, American POWs had assisted the British in the planning and digging of escape tunnels as well as with tunnel security. In addition, escapes were being planned from South Compound, although these plans were not as elaborate as those in North Compound.

The most well-known escape attempt was known as The Great Escape from North Compound. It occurred the night of March 24, 1944, after many months of planning. Seventy-six prisoners escaped through the tunnel Harry, which led to the outside of the camp into the woods. Of those seventy-six, only three were successful. Hitler was furious about the breakout and ordered that fifty of the escapees that had been captured, randomly selected, be executed. They were shot in the back of the head when they stopped by the side of the road to relieve themselves. The rest were returned to camp.

There are many excellent accounts in other publications of the tunnels, escape paraphernalia, and the Great Escape that detail the tragedy. I am sure my father knew some of the prisoners who were executed and was aware of the somber tone in the already dreary prison camp—not just for him but for everyone. In fact, the entire Stalag Luft III camp was allowed to have a memorial on April 7 for the slain POWs, once the British had completed a vault they had been permitted to build to hold fifty urns of ashes of the murdered men. The unspeakable deaths affected not only the prisoners but also the German guards and commandant.

Until then, it had been a prisoner's duty to escape, but that philosophy changed after the Great Escape and the executions. The British and American senior officers advised POWs that escape was no longer a duty of the officers. It was too close to the end of the war to risk escaping.

The German commandant was removed from the camp, much to the detriment of the prisoners. A new commandant, as retaliation for the escape, demanded that there be three appells a day instead of two. The prisoners cleverly maintained military order and discipline in the first two appells but were purposefully more disorganized and chaotic in the third. Due to frustration, the guards quit having the third appell after a while—just as the prisoners had intended.

There were other attempts by the goons to punish the prisoners in all the compounds. Restricting the use of the theater was one, and taking away musical instruments was another. There are also some accounts of the German food rations being cut, but the supplemental Red Cross parcels were sufficient to keep the Kriegies fed.

In the article "Stalag Luft III, The Secret War," in the anniversary brochure *The Longest Mission*, the author reflected on post-Great Escape repercussions: "It was still war. It was still prison, and it was still grim. With a madman on top, there was the ever-present threat that authority above the Luftwaffe could change things on a whim. Kriegies always knew that they were living on the razor's edge."

As the air war escalated in 1944, the number of POWs sent to Stalag Luft III increased dramatically. In addition, a new West Compound was opened April 27, 1944, with Colonel Darr H. Alkire in command. It was a totally American compound, and some of the senior officers from South Compound went over to help set up West Compound, which could hold two thousand men. Officers from Center and South compounds populated

West Compound, and new prisoners were funneled there as they arrived. With the prison population doubling between April and November, 250 prisoners were locked and shuttered in each cold and drafty barrack.

By mid-May 1944, about 3,500 Americans occupied three compounds—Center, South, and West—and the RAF prisoners still occupied North Compound and East Compound. Plus, some were sent to the off-site Belaria Compound with Americans. A photograph from the air of Stalag Luft III with an accompanying map are shown in Figure 8-1.

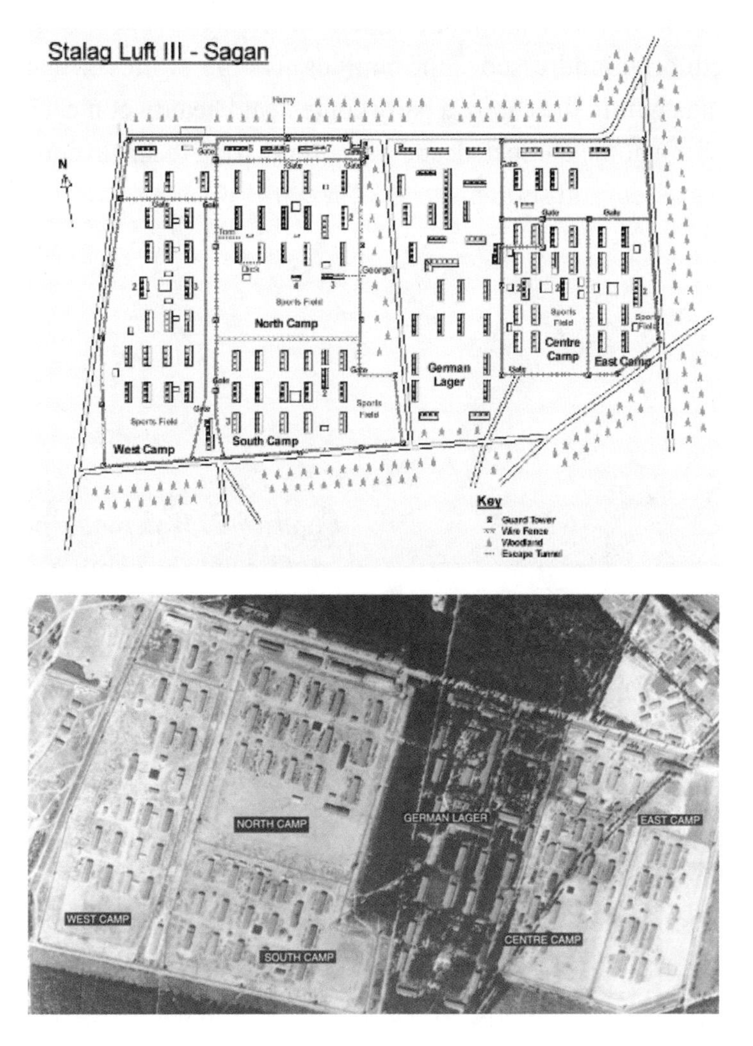

Figure 8-1: Drawing of Stalag Luft III with depiction of Great Escape tunnels (courtesy of pegasus.org) and September 1944 aerial view of Stalag Luft III (U.S. government reconnaissance photo)

Another morale booster for the prisoners in South Compound was the newspaper *The Circuit*, published three times a week and written and produced by the Kriegies. The German officers let the Kriegies use a typewriter with the understanding that none of the typewriter parts were to be used for escape purposes. *The Circuit* was a one-page newspaper with four columns and a hand-lettered title. It was posted on the cook-house wall, which was typically used as a bulletin board. Large crowds would assemble to read *The Circuit* when it was posted. There was also a newspaper in Center Compound called *The Kriegie Times*. Both papers carried editorials addressed to the families at home, stories on a variety of happenings within the camp, a baby contest and beauty contest in which pictures from home competed, and a section sharing recent events in camp theaters, classes, and sports.

Figure 8-2: Watercolor painting of Penny, the heroine of the "PENNY" cartoons by Ben Smotherman in The Circuit *newspaper. Her painting appeared in many Wartime Logs, including Daddy Ben's*

At the bottom of each issue of *The Circuit*, space was reserved for a cartoon strip called "PENNY," written and drawn by Ben Smotherman. Beginning in October 1943, Ben was delighted to create the cartoon strip, in addition to his incredible drawings and paintings in his Wartime Log. Producing those cartoons was another activity that kept my father's mind, hands, and creativity active. Penny's portrait (Figure 8-2) was painted into many a prisoner's logbook! A painting of Penny was also used as the grand prize in a bridge tournament. The size of each original comic strip was three and a half inches by thirteen inches, and ninety of them were drawn by Daddy Ben.

Ben's concept of Penny and his comic strip gained a lot of popularity among the prisoners. Today, Stalag Luft III historians and authors have made "PENNY" famous, and my father is more well known for being the artist and creator of "PENNY" than probably anything else he did.

As my father told the story:

If I got nothing else from Kriegie life, I did get some good experience in cartooning and watercolor sketching. Most of my spare time (and there was plenty!) was taken up in some sort of drawing. It was divided between "PENNY" and logbooks. I don't know how many drawings I've done in logbooks, but I'm sure they number in the hundreds. I did many types of drawings in them, but the majority were "PENNY" girls in various revealing poses.

"PENNY" was originated to satisfy a demand. In September 1943, shortly after we moved into this [South Compound] *from Center Compound, Ollie Chiesl, Roy Deadman, L.J. Hawley, and several other literary minded Kriegies originated* The Circuit*—our South Compound newspaper. Roy knew of my ambition to be a cartoonist, so he asked me if I'd like to draw a comic strip for* The Circuit, *then unnamed. "Something about a girl" was the only specification he gave me. So, when* The Circuit *made its first appearance October 6, 1943, my strip was tacked on the bottom.*

I don't know how I came to name the girl "Penny." I wanted a name that was short and unusual, easily pronounced, and, if possible, catchy. "PENNY" came to me after a short session of concentration.

As for the appearance and character of Penny, that wasn't so hard. Of course, she would have to be a blue-eyed blonde! To avoid a reputation that would have been very easily established, I made her a "nice" girl and have kept her such. Penny is a girl who knows what men think about and doesn't let them put thoughts into words or actions!

Next, to fit Penny in with the thoughts of Kriegies, I had to have her associated with the war somehow. This wasn't hard either. I made her a war correspondent in England. Here, she was close to the war yet remote enough to be interesting to a prisoner of war—my public.

After a few introductory strips, Penny went on a raid to France and was shot down. After several months of evasion, she and "Sunshine," the pilot, made their way back to England. Then came the invasion of France (my timing was a little off, and I had to fill in a few build-up strips!) and she made a trip to the invasion coast. While she was there, I learned something. A POW came up to me one day and said, "When is Miss Penny going back to England? We don't want anything about war—we want sex!" I promptly hurried her back to England and sent her on a vacation to the beach! Of course, I can't satisfy all requests—it wouldn't suit Penny's "character"!...

I suppose "PENNY" has fulfilled its purpose. It is still read after a year's running, though I admit the competition isn't very keen. At least it has been helpful to me, and I think it has to others as well.

THE CIRCUIT

RELIGIOUS SERVICE HELD IN THEATRE

EDITORIAL

CAMP ORGAN MAKES DEBUT OCT. 6, 1943

FOUR FILMS SHOWN SINCE SEPTEMBER

KRIEGIE LYRICAL NEEDS SUPPLIED IN VARIED FORMS

STAFF

PENNY

By Ben Smotherman

Figure 8-3: Ben Smotherman's "PENNY" cartoon strip in South Compound newspaper, The Circuit *(Red Cross Prisoners of War Bulletin, August 1944)*

I don't know how much my mother knew about the "PENNY" comic strip, but I'm sure my father told her about Penny and how her likeness strongly resembled my mother's! My mother was an attractive blue-eyed blonde whose hairstyle and appearance looked very much like Penny's! A sample of *The Circuit* newspaper, including the "PENNY" comic strip, is duplicated in Figure 8-3. This version of *The Circuit* appeared in a Red Cross bulletin that was sent to POW families.

As spring 1944 approached and temperatures became milder, the prisoners were able to participate in sports outside. In addition to the sports already mentioned, fencing was added, completing the sports fields.

Warm weather encouraged the creation of model ships to float in the fire pools. At the same time, prisoners began building model airplanes. As my father was an avid wood carver and airplane aficionado, I can't help but think that he was involved in that pastime!

My mother, Ben's parents, and Ben's sister rotated the opportunity to send parcels to him every sixty days. Ben exclaimed in a card dated March 22, 1944, that he had received his second parcel and more books. My mother's next opportunity to send a package was May 29, 1944. Ben's requests were similar to those he had for the earlier package that my mom sent—hygiene goods, a few clothes, and a pipe!

Most of the letters my father wrote to his family in 1944 had recurring themes: "I'm fine—don't worry about me," "I'm healthy and getting exercise," "I'm staying busy with my artwork," "take care of yourselves," "send photographs," "getting plenty of rest," "getting closer to coming home," "fried chicken," "can't wait to see everyone," "thanks for the letters—they mean so much," etc.

In a letter from my mother to her brother-in-law dated April 10 (Ben's birthday), which thanked him for sending a birthday present to mail to Ben, she said: "This past week, I received one card and two letters from Ben. The card and one letter were dated February 1, and the other letter was dated January 16. They were all very nice and sweet—however, they didn't say much. He said that he received his parcel and it was very nice. He asked for photographs and photographs and photographs. Most of the mail was just saying 'sweet nothings.'" I'm sure that Ben was aware of the censors reading his cards and letters and did not want to include much information about the camp or being a POW.

At the same time, life at home for my mother had changed. By April 1944, my mother started working again. Based on the letterhead that she used in the April 10 letter, it seems she was working as a U.S. civil servant for the Tenth U.S. Civil Service Region (comprising Louisiana, Mississippi, and Texas) in Dallas for the Office of the Director. I believe that she was a secretary or some sort of executive assistant. When I became a U.S. civil servant in 1979, my mother spoke fondly about her time in that role and indicated how much it had helped her stay busy while my father was a POW. I believe that she used her time well by working, cooking, sewing, reading, and studying at Southern Methodist University. I don't know how she did it, but she also remained physically active by learning to play tennis and golf and getting involved with bowling.

Just two months later, both of my parents' lives were impacted when Allied forces finally broke through to Europe. The POWs knew it as "The Invasion," and they spoke in hushed tones about it for a long time.

One night, during the monitoring of the nighttime communique from BBC, they heard code words declaring the imminent invasion on the coast of France. All the Allied forces—ground, naval, and air—had a role in the invasion. The purpose of the bombings leading up to the invasion was to gain the air superiority needed to ensure a successful invasion. The Allies had continued the bombing campaign (Operation Pointblank) of which my father had been a part, targeting German aircraft production and airfields and adding targets of fuel supplies and railways. In addition, the USAAF was called upon to change some of their strategic bombing to support the tactical movements of the soldiers on the ground.

On June 6, 1944, at about 10 a.m., prisoners who spoke German gathered daily to listen at the loudspeakers mounted in every compound, as they normally did. That day, the broadcast started routinely, but then there was a pause, and the invasion in Normandy, France, was announced in fairly benign terms. But the POWs could not be fooled; they knew the significance of what they were hearing. The entire camp of 10,000 men erupted as the news spread in minutes. My father's drawing of the announcement is in his Wartime Log (Figure 8-4) and reflected the mood of the camp that day. The place was wild with excitement! All felt that if the Allied forces had made landfall in occupied France, then the war

Figure 8-4: Drawings from Ben Smotherman's Wartime Log

was coming to an end, and they would be going home. In a card to his family dated June 10, 1944 (received September 1), Ben was careful not to mention the "event" by name, fearing censorship:

Dearest Folks, How's tricks? Guess you were as glad about this month's news as I was! Let's hope I'll be seeing you soon and helping do away with some of that fried chicken! I'm still feeling fine, in good health, and hope all of you are the same! I hope Don sticks around [the U.S.] *until I get back. Take it easy, and don't work too hard—it isn't healthy!*

The POWs were obviously hopeful that the war would be over soon because of the invasion of Normandy, but it was not to be. However, the invasion of Normandy and the increasing American air superiority turned the direction of the war around in mid-1944, and the Allies were then winning.

During the summer of 1944, the Germans increased their harassment of the prisoners. They were unsettled because Germany was losing the war. According to Bub Clark's book, the guards placed limits on cigarettes and the amount of Red Cross food that could be stashed. It was difficult to have an emergency stash of food anyway because the Germans punched each can so that the contents had to be used right away. They did not want the prisoners keeping emergency food stores for escape attempts. In addition, the guards initiated arbitrary but endless special appells.

On June 16, my father received his third parcel, so that helped his morale! He saved labels from items in the parcel, namely Garden Vegetable Soup-er-mix, Woodbury facial soap, and Keen's white pepper. He also saved labels from two Vassar union suits (long johns), which were 75 percent cotton and 25 percent wool, made by the famous company in Chicago. Once again, he let people know how much he appreciated their thoughtfulness.

By July 4, the prisoners' hopes of victory had waned, but they had a great Fourth of July celebration, even inviting the British to take part. At the end of the day, due to consumption of their homemade brew, many men did not make appell, and some top officers, including Clark, were thrown into the fire pool! Ben wrote in a letter to his family dated July 9, 1944 (received December 11):

Dearest Folks, I received a pile of mail on the second, and my morale really rose! ...I'm glad all of you are well again and hope you remain so. I'm still fine. We had quite a 4th here—mostly sport events—and a gay carnival. I hope to see you soon, so keep the home fires burning!

Portions of letters from Ben to my mother were later typed by his family. In a letter dated July 23, he acknowledged receipt of the March-May parcel, calling it "perfect!" According to him, everything was in good shape and fit perfectly. He later wrote on my mom's birthday, on October 24, 1944:

This is your day, angel, and I am thinking especially of you and hoping you are having as happy a birthday as circumstances permit... I'd better wish you a merry Christmas and a happy New Year too! I think 1945 will be our happy year! I'm so glad you are doing so well in school, honey—I'm mighty proud of you... I've received fourteen of your letters this month, with six pictures, so my morale is high!

July 28, 1944, was the one-year anniversary of my father's confinement. As he put it in an expression on a drawing in his Wartime Log (Figure 8-4), he "started his sophomore year in Sagan University." On that same page, he illustrated on June 17, 1944, "I gave my Spam to some maggots!" So, living conditions and food were still an issue.

By September, due to transportation and supply issues, the parcels allocated to each prisoner dropped to about one parcel every two weeks. The camp was again overcrowded, so in addition to the triple bunk beds, big white tents were erected to accommodate the growing population. By then, South Compound had six full colonels and seven lieutenant colonels among the two thousand Kriegies.

Fall 1944 came, and the war was still raging. Several accounts by the POWs said it was their lowest point while in the camp. Faced with another winter in the prison and another Thanksgiving and Christmas away from family, prisoners certainly had a tough time. Plus, with the reduced Red Cross parcels, the overcrowding, the unpredictable behavior of the guards who could be "trigger happy" in the goon boxes, and the heavy cloud of gloom hanging over everyone ever since the murders after the Great Escape, it was hard to raise morale. My father often wrote that he was anxious to see everyone and hoped that it would be soon.

On September 30, he wrote:

Dearest folks—I hope this finds you all well and kicking around! I'm still fine and healthy. Time is still passing quickly in spite of a wave of optimism. I guess it's because I keep busy. I'm still drawing my cartoon strip and gash (Kriegie jargon for "extra") stuff! I still hope to see you soon (been hoping fourteen months now).

Thankfully, the Thanksgiving parcel from the International Red Cross was extra special in 1944. From the parcel, my father and his roommates from barrack 133, room three, were able to put together a Thanksgiving meal and created a hand-drawn menu (as shown in my father's Wartime Log, Figure 8-5).

He further told the story of that special day, including an effort by the prisoners to make the best of the situation and lift spirits:

When a Kriegie looks back on his "time" in Stalag Luft III, he finds few days that stand out conspicuously. Time passed in lumps of a week or a month, and there weren't many days exciting enough or unusual enough to merit a groove on the prisoner's memory. There were some few, however, that we all remember—happiness we found for a day—and these we will never forget.

Figure 8-5: Thanksgiving 1944 menu, as drawn by Ben Smotherman in his Wartime Log

Such a day was Thanksgiving Day, November 30, 1944. For a day, we celebrated, and our day-to-day schedule of existence was disrupted. Everything possible was done to take our minds off the barbed wire and guard boxes. Plans were laid out several days in advance.

Perhaps the most unusual was the food. Ordinarily, we have little variety in the culinary department from week to week. This day, however, Ralph dreamed up a dish never before tried in a Kriegie camp. Aided and abetted by Joe, Ralph set before us baked potatoes, Romany style! They removed a cylindrical core from the potatoes (with an implement made by Joe from a tin can), stuffed them with jam, and plugged the holes. Unanimous comment made by the critical roommates: "Um-m-m good!"

Supplementing the baked potatoes, Romany style, were biscuits by Van, fruitcake by Dave (he also prepared the grilled cheese appetizers), and other dishes worthy of a place in Kriegie history. Parker helped paste up the meat pie, and odd jobs were handled efficiently by others. Needless to say, our hunger was completely satisfied—but prematurely! After the dinner, odd bits of biscuit, cake, etc., could be seen resting in ignored splendor on various shelves while on the bunks below them one of us lay stretched in blissful repose. To say that we had plenty to eat would be putting it mildly!

Sharing the honors with the dinner, the phonograph played an important part in the day's activities. It was loaned to us by Jerry Scollard of room ten, and records were loaned by Fryer, Des Hotels, Barnes, Moritz, and La Chasse. Dave sharpened some used phonograph needles, and Frank acted as record changer. Music ran from symphony to swing to "slow and sweet"—and back again. Roy dug out some choice letters from Claire to help him along, and we all worked on a brief (?) case of homesickness. To the music of the All-American Bands, we dreamingly gazed at pictures of our loved ones, and I'm sure there were fourteen silent prayers—not only of Thanksgiving but for the future as well.

For this, our day of Thanksgiving, we donned our best uniforms, and ties were the order of the day. Visitors dropping in wondered where we were going and, when told that we were merely observing Thanksgiving, said "Oh" in an all-knowing way.

It was truly a day that we'll all remember, though it will sound dull to one who is accustomed to the varied life of civilization. To us, it was something different, and anything different from our monotonous life was a blessing to us. We went to bed that night feeling a little better, though more homesick, after this reminder that better things were to come. Before drifting into care-removing slumber, another silent prayer was offered. "God grant that we shall be home next Thanksgiving."

But, alternatively, Christmas 1944 was a melancholy one, as the POWs had thought that the war would be over by then and they would be liberated. There was no mention of Christmas at all in my father's Wartime Log. Nevertheless, the Red Cross parcels helped to make the season more festive, and much planning occurred to make "turkey and the fixins." There was plum pudding, turkey (boned meat), small sausages, strawberry jam, assorted candy, deviled ham, cheddar cheese, mixed nuts, bouillon cubes, fruit bars, dates, canned cherries, playing cards, chewing gum, butter, assorted games, cigarettes, smoking tobacco, pipe, tea, honey, washcloth, and pictures of American scenes. A new transportation scheme was devised whereby the parcels were shipped by sea via Sweden to a port in Germany near the prisons, as the German railway distribution had been severely damaged by bombing.

To drive the prisoners further into the dumps, it was learned that on December 24, 1944, the German offensive in the Ardennes, the Battle of the Bulge, was successful. In prisoner Edgar Denton's terms, "It was like somebody was stealing Christmas." Others conjectured that the Germans knew that they were losing, and this was their last gasp of breath.

A repatriated POW from the camp was interviewed in the last part of February and said: "Yes, I know Ben Smotherman—not personally, but I passed him every day and knew who he was. He is in good health and is getting along alright. Ben, you know, has won great fame in the camp through his artistry. Everyone knows Ben."

Meanwhile, Ben's brother, Don, graduated from pilot training and was assigned as a B-17 pilot at San Angelo, Texas. Ben wrote a card to him, saying:

Dear Don, I can't tell you how glad I was to hear of your graduation, though I never doubted for a minute that you'd finish. Try to be satisfied with your job, and don't try to go glory hunting. I'm proud of you, boy! I hope, of course, that you get a promotion soon, as you certainly deserve it. Keep your head out! Ben

In March 1945, Don was ready to go to war, but he never left. Bill Stewart's fears were allayed. By then, Bill had flown both Schweinfurt raids that were very difficult for the 8th Air Force, and heavy losses were experienced. However, he survived those ordeals, completed his twenty-five missions, and went home to Texas. Things were looking up.

Chapter Nine
The Long Winter March

The new year of 1945 brought both hope and apprehension. Their motto for 1945 was "'45 and Alive!" Ever since the execution of "The 50," rumors circulated that Hitler wanted to kill all of them. It was also well known that the Gestapo and the Schutzstaffel (SS) were liquidating prisoners at other prison camps. That instilled apprehension in the very core of the hearts of the POWs.

In mid-January 1945, the Russians reached the Oder River (about fifty miles northeast of the camp), and the prisoners could hear the planes and artillery that were approaching. Thanks to secretly listening to BBC radio, the prisoners knew it was the Russian Allied forces they heard. That inspired hope. But the prisoners worried about what the Germans' plan was for them if the Russians got close to the camp. Would they all be killed? Would they be abandoned with no way to get food or supplies for 10,000 men?

About a month before that, the SAOs had discovered from their intelligence sources that, with Allied troops encroaching upon the area, there was a strong possibility the Germans would either evacuate the prison camp and let the Russians take over or would make the almost eleven thousand prisoners march in the dead of winter to another prison camp.

Therefore, in January 1945, the SAOs required the American POWs to run or walk ten loops around the "circuit" (one-quarter-mile perimeter) per day in order to get in shape for a long walk. The weather and conditions were not ideal, and the groans and complaints from the prisoners grew by the day. According to Bub Clark, the prisoners were also instructed to sew a hood onto their sweaters or jackets and to make sure they had good shoes, a wool cap, a prepared pack, two pairs of long johns, and two pairs of wool socks handy. They were also told to stash as much food as they could, especially chocolate and cigarettes for bartering.

In addition, the British and American senior officers decided that a fresh batch of Red Cross parcels received in January could be released to the prisoners rather than being stored. The senior officers thought that the prisoners should eat three hot meals a day and "fatten up" for a potential march.

On January 27, 1945, at Hitler's 4:30 p.m. staff meeting in Berlin, he issued the order to evacuate Stalag Luft III. He was fearful that the Allied POWs would be liberated by the Russians, and Hitler wanted to keep the POWs as hostages. The Soviet Union's southern army was already within twenty miles of the camp by then.

That Saturday evening, many Kriegies were watching the Kaufman and Hart play *You Can't Take It With You* in the South Compound theater. Making an unscripted entrance, SAO Colonel Charles G. Goodrich strode onto center stage and announced, "The goons have just given us thirty minutes to be at the front gate! Get your stuff together and line up!" Urgency was dictated by the Germans' desire to evacuate in plenty of time before the Russian troops arrived.

In the barracks following the sudden and dramatic announcement, there was a frenzy of preparation. Sleds were improvised using the wooden boxes in which the Red Cross parcels were shipped. More sleds were created with overturned benches that had runners attached to them. Large amounts of food were consumed—to the point some men became sick— as it could not be carried.

In addition, backpacks were made by tying the legs of pants or the arms of shirts together into a knot and placing them over the head. In these kluged together sacks, the essentials—stashed food, minimal clothes, and personal items—were packed. Most men abandoned such items as books, letters, camp records, and even their treasured Wartime Logs but took their overcoats and a blanket.

Thankfully, my father packed his Wartime Log and a rolled-up bundle of comic strips of "PENNY." He also carried the first letter he received in prison camp from my mother.

My father's Wartime Log reflected the chaos as everyone got ready to evacuate:

"They'll never evacuate us!" We repeated this phrase mockingly as we rushed about in hasty preparation for the night march. Everybody was in a flap. A few minutes had elapsed since the colonel [had] *marched down the hall, loudly announcing, "Prepare to evacuate in thirty minutes!" and we were now rushing about like disturbed hornets, collecting the things we had previously decided to take along. The cupboard was almost thrown open and roughly divided into fourteen piles on the table. Discarded clothes flew through the air to fall limp in a chair or on the floor. Footsteps resounded in the hall, and locker doors banged. Shouts rang through the barracks, and confusion reigned. "Are you going to take these handkerchiefs?" "No, take all you want!" Old clothes were stripped off to be replaced by "Sunday best"—it would be better to wear them than carry them. It was nine o'clock.*

Figure 9-1: Ben Smotherman drawing of hectic thirty minutes prior to the long winter march, from his Wartime Log

A picture that my father drew of this hectic event is shown in Figure 9-1. Some of the prisoners delayed on purpose, hoping that the Russians would take over. Others feared that the Germans might take them hostage or shoot them before the Russians arrived. Most of the pandemonium was a result of lack of preparation and organization on the Germans' part. Hitler changed his mind several times about the evacuation and waited until the very last minute to order it. Despite the POWs' efforts, much had to be left behind—sports equipment, musical instruments, books, and Red Cross parcels.

My father continued:

As I passed by Padre's room on the way outside, I looked at my watch in the light streaming through the open door. Nine-thirty. We had packed and were ready in thirty minutes. Thirty minutes of confusion! Outside it was cold. Men were standing around in groups, talking and gesturing excitedly, waiting to form in orderly ranks. Minutes went by, and the packs began to get a little heavy. Benches and stools were brought from the disordered rooms, packs were eased to the ground, and we began to relax a little from the nervous tension that had gripped us.

The men stood outside their barracks in six inches of snow for two hours, and the snow was still falling. There have been various conflicting reports of the temperature, but it was sixteen degrees Fahrenheit with a stiff wind. Once on the move, each prisoner was given a Red Cross parcel at the gate. Ugh! Something else to carry! The German guards with their sentinel dogs herded the men through the main gate and warned that anyone who lagged behind would be shot. Outside the camp, Kriegies were counted and waited outside as the icy winds penetrated their multi-layered clothes and froze stiff the shoes on their feet.

From that point on, my father kept another small journal, as his precious Wartime Log was packed away. In the new journal, he recorded leaving his barrack at 9:30 p.m. Finally, the two thousand men from South Compound were the first to leave Stalag Luft III at 11 p.m. in the bitter cold and snow. Struggling against the icy wind and exhausted from the late-night march, the column of men, three wide, led the way, with over eight thousand men behind them. After South Compound left, West Compound followed, leaving about 12:30 a.m. At 3:45 a.m., North and

Center compounds departed, and at 6 a.m., East Compound pulled up the rear. By dawn, the line of ten thousand marching prisoners stretched twenty miles.

There were about five hundred men who were too weak to walk, and they were looked after by the medical staff. Fortunately, there was enough food left behind that they had plenty to eat. Eventually, on February 6, the injured and weakened men were loaded onto box cars and transported to Nuremburg, arriving on February 10.

The marching prisoners did not know where they were going, and they were scared as well as very, very cold. The guards from the camp, older and many unfit, marched alongside the POWs, carrying rifles and machine pistols. The entire massive group continued all night, taking ten-minute breaks every hour.

According to Ben:

Slowly the men began filing through the gate, hastily counted by Germans who were just as excited as we were. Stop and go, stop and go—it began to get tiresome. It was eleven o'clock before we settled down to a reasonably steady march.

"Give way to the right!" The cry passed quickly up the long column, far ahead of the heavily loaded truck that had occasioned it. A motorcycle chugged up the side of the column, emitting strong gasoline fumes. Out of the darkness loomed a wagon pulled by steaming horses—a load of civilian evacuees.

The column halted at about 2 a.m. [January 28]. *Word was passed down the column that we would be there in about an hour. Fires were kindled in the road, fed by discarded clothes, logbooks, and other items from the packs. We were issued bread—frozen German bread. The warmth created by marching soon wore off in the biting wind. We were on a high hill in the road and completely exposed to the wind. Everybody was eating a snack and trying to keep warm.*

Ben's drawing of the men by a fire during this stop is shown in Figure 9-2. The first stop of any significance after sixteen miles was in Kunau, Germany. It was there where the weary POWs built fires and threw into it precious logbooks, sacred letters, and also treasured artwork in order to keep warm. Pictures of this long winter march were taken by clandestine cameras, and one is shown in Figure 9-3.

Figure 9-2: Ben Smotherman drawing of POWs taking a break and trying to get warm with a fire during the long winter march, from his Wartime Log

After about two hours, the march was resumed. How stiff we were! The excitement was gone. We no longer jokingly shouted to each other. This was no joke. Our feet were sore after the long halt and cold. The packs grew heavier and heavier, and we began to wonder what else we could discard to lighten them. Logbooks lined the side of the ice- and snow-covered road. A guitar case was seen leaning against a tree. The march was beginning to tell on us. Men began to lag, to fall out, to be helped along by their comrades or be placed on the crowded wagon at the end of the column. We began to realize the seriousness of the situation.

In the east, it was turning gray. Dawn was coming, bringing with it, we hoped, a rise in temperature. The sky was overcast, and we saw no sun, but the light itself seemed to cheer us up. Looking ahead and behind, we could discern the slow-moving column, stretching like a dark ribbon across the dreary white landscape. Ahead, we could see it turning to the right; behind, it bent in a gentle curve, silhouetted in the white snow and ice.

Figure 9-3: South Compound on the long winter march; photo taken with POW clandestine cameras (photo courtesy of Clark Special Collections Branch of the USAF Academy Library)

Figure 9-4: Ben Smotherman drawing of the column of men on the long winter march, from his Wartime Log

Figure 9-5: Long column of men during the winter march; photo taken with POW clandestine cameras (photo courtesy of Lt. Harold Kious)

Ben's drawing of this column of prisoners is shown in Figure 9-4, and a photo taken with clandestine cameras is shown in Figure 9-5. The path along the march was littered with items that the Kriegies wanted to take with them but could no longer carry. Musical instruments, models of boats and airplanes, books, a straw-stuffed mattress, a typewriter, and even food and clothing were strewn in the snow. The men continued to Grosselten where they sought cover in barns and stables and slept on hay. In the morning, they ate and got hot water from the farmer so they could make coffee.

In his journal, my father continued:

At 11 a.m., January 28th, we stopped at Grosselten. We crowded into barns and found some relief in a brief sleep and some food and hot coffee. Packs were further lightened. I had thrown away nothing so far, but here I discarded a suit of winter underwear, a bath towel, and a carton of pipe tobacco. These I could do without, and the weight, little as it was, would make a difference.

The Kriegies were dressed in every conceivable kind of attire. I was wearing winter underwear, a pair of pink pants [the USAAF uniform pants], *a pair of khaki pants, a green shirt and two khaki ones, a muffler,* [and] *a wool cap, and I had a face towel wrapped around my head and ears. I wore a pair of light socks and a pair of woolen ones and G.I. shoes. Add to this my RAF coat and leather gloves, and I was dressed for marching. My pack was one I had made from material from an English duffle bag and was reasonably comfortable. I think it weighed not more than twenty-five pounds, but at times, it felt three times that!*

About 4 p.m., we started leaving Grosselten. Again, we went through starting and stopping, marching ahead, marching back, standing around, and so on for over an hour. We were issued some margarine. At last, we were marching steadily again. Every fifty minutes, we stopped for a ten-minute rest, which served more to stiffen our limbs and numb our feet than to relieve our weariness. After such a stop, it took about ten or fifteen minutes [of] *marching to warm our feet and loosen our knees.*

Figure 9-6: Ben Smotherman drawing of men taking a rest break during the long winter march, from his Wartime Log

Figure 9-7: POWs taking a rest break during the long winter march (photo courtesy of Clark Special Collections Branch of the USAF Academy Library)

Ben's drawing from his Wartime Log of one of the rest stops is shown in Figure 9-6, and a photograph of some Kriegies resting is shown in Figure 9-7. The men of South Compound set the pace for the entire marching column, and frostbite and exhaustion took their toll. When a man went down, and he felt he could no longer go on, his buddies carried his pack and held him up to march along until he was better. What do you suppose was on their minds? They considered the questions: Where are they taking us? How long will we have to march? When do we eat? When do we sleep? Will we survive? Ben continued:

Toward dusk, it began to snow, and the temperature dropped. A stiff wind drove the powdery flakes into our faces. It was difficult to breathe. The temperature dropped lower and lower, and our spirits dropped with it. Sore feet made us wince at each step, and the column began to string out more and more. Vainly we tried to keep the gaps closed up. Men dropped out for a brief rest on a stump, others to be picked up by the end of the column. We were headed for Muskau.

A road forked off to the left, and we took it. Where were they leading us? Thoughts of dropping by the roadside and freezing brought on a little more strength and effort, and we trudged slowly on. We had started from Grosselten at a rather brisk pace and were now barely crawling along. How weird the men looked dragging along stiff-legged and numb.

The march was harrowing for everyone but was hardest on the men from South Compound, who were at the head of the line and were berated and hurried along by the German authorities. Russian guns could still be heard in the distance, and the lead compound had to move rapidly. They marched over thirty-four miles from Sagan to Muskau in twenty-seven hours with only four hours of sleep. This was in clear violation of the Geneva Conventions, which specified that POWs were not to be marched more than twelve and a half miles in twenty-four hours. Rations consisted only of margarine carried in coat pockets, where it froze, and brown bread carried on horse-drawn wagons until it ran out.

Arriving in the industrial town of Muskau at 1 a.m. on January 29th, they had to stand out in the cold for two hours before it was decided to put all of them inside. The following is a description by my father of that hike:

The eighteen kilometers between Grosselten and Muskau turned out to be thirty, and we dragged into a glass factory at 2 a.m. on the 29th, stiff, numb, and half frozen. There were fires in the glass factory, and we crowded in eagerly. There wasn't room for us all. We pressed together to make more room. Still, men were standing in the freezing cold outside. A brick factory was opened to make room for the others. Somehow or other, we all got inside out of the weather and began to prepare for the night...

It turned out that the large factory in Muskau made ceramics, glass, and earthenware. There were several buildings in the complex, all providing shelter and warmth as well as running water. Men huddled around the warm glass furnaces after being in the bitter cold. Their feet were treated for frostbite and blisters by prisoners who administered first aid. The total time spent in Muskau by the prisoners from South Compound was thirty hours. There were many stories of the kindness and generosity shown by the German townspeople of Muskau, who offered food and shelter.

The march by foot of South Compound from Stalag Luft III in Sagan to Spremberg was hand-drawn by Bub Clark, then redrawn and included in a map in Figure IV of the appendix.

My father's logbook provided more details:

All the next day, we rested, prepared hot meals, and discarded more weight from our packs. More sleds were made... We had successfully passed the low point of the trek, and no men had been lost.

About fifty men from South Compound stayed behind and joined the West Compound prisoners who followed them to the glass factory because they were the victims of frostbite or continual dysentery, which plagued many men on the march. The routes of the men from North, East, and Center compounds varied from time to time as the Germans experimented with finding shortcuts. In the beginning, even the Germans did not know where they were going.

My father continued:

Tuesday morning [January 30], *we left Muskau, leaving a few men behind who were unable to march. We had greatly recovered from our weariness, and prospects of a short eighteen kilometers to Graustein left our spirits high. It was much easier pulling our packs on a sled, and Roy and I alternated in thirty-minute periods.*

By 7 p.m., we were in a barn at Graustein and were soon settled comfortably in the hay. After a brief snack, we literally "hit the hay" for a good night's rest.

Next day [January 31], *we were on the move again, about 8:30 a.m. After a short march of only seven kilometers, we arrived at Spremberg and were put in a large garage. Here, we were given hot barley, meat, and potato soup, which was very good and very refreshing.*

The weather had warmed up, and the POWs had to deal with mud and slush as the snow melted, and many discarded the sleds that had served them so well. The distance from Muskau to Spremberg was fifteen and a half miles. By the time the prisoners reached Spremberg, they had spent three days and three nights in transit and had covered fifty-two miles. The POWs there were taken to a large military garage at a German tank facility, and after eating hot soup, they were moved to military barracks until they were marched out to the train station in Spremberg. Ben's accommodations in the large garage are shown in Figure 9-8.

Figure 9-8: Ben Smotherman drawing of POWs arriving at Spremberg and being housed in a large garage, from his Wartime Log

In Spremberg, there was a long line of WWI era box cars (Figure 9-9), which were ready for the prisoners to board. They had recently been used to haul livestock and had not been swept out, so the smell and filth were disgusting. Each car was known as a "forty and eight," meaning forty men or eight livestock would comfortably fit. For South Compound, there were two thousand prisoners who would be locked up in twenty-eight to thirty cars, so there would be fifty to sixty prisoners per car. There were so many men that they could not all sit or lie down. If they sat, it was in a line in a "toboggan" fashion. Otherwise, half sat down while half stood. Colonel Goodrich and other senior officers found a car in the middle of the train. The goons occupied the front and rear cars and had sentinels on top of those cars during the loading operation. No one knew where they were going, including the guards.

Figure 9-9: Railroad box cars in Spremberg waiting to take POWs to Moosburg (photo courtesy of Clark Special Collections Branch of the USAF Academy Library)

News that we would board a train here and march no more left us happy. We left for the train about 4:00 p.m., and after a brief march, we were piled in a boarded-up cattle car just before dark. The floors were wet, and we were crowded, but we thought it better than marching in the cold. The car measured about ten feet by thirty, and there were fifty-two of us aboard.

After a rattling journey through rolling hills and bombed marshalling yards, we arrived at Stalag VIIA at Moosburg, Friday, February 2, 1945.

Travel by train took two days and three nights, and the trains stopped about every twelve hours so prisoners could take a break and get a drink of water. Most men were sick with colds or dysentery. The railcars became engulfed with urine, feces, and vomit. The only ventilation in some of the cars came from two small windows near the ceiling on opposite ends; others only had ventilation through cracks in the wooden sides. The conditions were awful, but my father could not write in his logbook about them because the censors could confiscate the book. Then, all would be lost.

Word was passed along to the prisoners by Colonel Goodrich at Regensburg, that they were authorized to escape if they wanted to. Toward the end of the journey, thirty-two POWs managed to escape, but all were recaptured within thirty-six hours and returned to their groups.

Once again, the Geneva Conventions were violated, as the railcars were supposed to be labeled "POW" if they were transporting POWs. However, the railcars were not labeled appropriately.

In addition to the men in the box cars traveling under horrid conditions, they often wondered if they would be bombed by their own forces since train stations and railways were primary targets of the Allied bombers. For many of the prisoners, the days spent in the crammed box cars were the worst they had to endure, and that included the cold march.

Thirty-six hours and four hundred miles after leaving Spremberg, the train full of prisoners pulled into Moosburg, thirty miles northeast of Munich. A map of the long winter march that my father took is shown in Figure IV of the appendix.

Although my father arrived in Moosburg at 10 p.m. on Friday, February 2, the fetid box cars were not opened until 8 a.m. once the Germans found a place to put them. Banging on the box car doors could be heard throughout the night.

Upon leaving the cattle cars, the men were hastily organized into an appell and counted. Then, men were searched, and some Wartime Logs were confiscated or at least had pages torn from them. All the prisoners' clothing and gear were taken from them, and they were herded into big rooms with showers. I'm sure that the POWs had heard of the concentration camps with Jewish prisoners where many were herded into a room to supposedly take a shower and then were gassed to death. I know that the POWs were worried that would happen to them. However, they had brief, warm showers while their clothes and belongings were fumigated. It was the first hot shower they'd had in a month and a half, as Stalag Luft III only had cold showers in the last months.

The evacuation from Stalag Luft III was over, and displaced prisoners had gone to various prison camps across Germany. Unfortunately, these camps were worse than the accommodations and lifestyle at Stalag Luft III. The men had marched through one of the most bitterly cold winters in Germany, gone with little to eat, experienced extreme exhaustion, and been piled into filthy and cramped conditions in the box cars. Some were diseased with infected stomachs, dysentery, colds, pneumonia, and frostbite.

After arriving at Stalag VIIA, Ben was initially housed in a temporary "lager" in a filthy stable, which was extremely crowded. Two days later, on February 5, he was deloused and moved to barrack VB, which was also crowded. On February 6, he was moved to barrack VIIA, and on February 7, he was moved to barrack VIIB, where he was in a six-man combine with Andy, Jo, Boots, Cy Morrison, and Dave Nash. They were elated to discover after their move to barrack VIIB that they were the recipients of three Red Cross parcels! That was the first good food that they'd had since the long winter march had begun.

On February 8, Ben made some slippers out of the tops of some flying boots. He also made a watchband, probably out of scraps from the same footwear. He also noted that spirits were much higher with the receipt of the Red Cross parcels, as they had just received their fourth and fifth.

February 9 was a momentous date, as my father experienced his first flea bite! After all that he had been through, he kept his sense of humor, noting, "Big rascal! The fleas seem to like fair complexioned fellows, for they bite them most, it seems."

Things must have settled down by Sunday, February 11, as my father was able to attend church services at the prison camp. He attended church regularly before and after the war and read his Bible constantly.

Also on that Sunday, the British sergeants gave a band concert, but my father decided to read instead, as the British had loaned the Americans several books. Conditions there were terrible, but there were some ways to relax and forget about where they were living.

Of all days, on February 14, Ben temporarily moved to a crowded barrack so that his barrack could be fumigated. Things were looking up though, as he received his seventh Red Cross parcel and wrote to my mother. After the fumigation, Ben was able to move back into his barrack, attend church services, and receive another Red Cross parcel, so things were as normal as they could be.

Ben underwent tremendous suffering, malnutrition, and cramped quarters during the long march, so I always wondered how he managed to march and be stuffed into box cars while making it a priority to carry his Wartime Log and the roll of "PENNY" comic strips. During the march, he kept notes and dates in a small notebook that he could easily carry. After he arrived in Stalag VIIA, he retrieved his Wartime Log and then documented the long winter march with drawings and writings. The writing style changed after he reached Stalag VIIA, and his writings and drawings were in a heavier ink or a thicker pencil. Perhaps that was all he had available to him there.

News about the marches and subsequent trips to various prison camps began to reach the American and Allied hierarchy in mid-February. It was not clear to anyone why the Germans had evacuated the prison camps as opposed to leaving the prisoners there to be rescued by the Russians. The German propaganda machine tried to explain that the POWs had volunteered to march because they did not want to be released to the Russians.

As the numbers swelled to around 130,000 in Stalag VIIA by April 1945, the families of the prisoners had very sketchy information and expected the worst. They did not know where their loved ones were, what their condition was, or how they were being treated. I don't know what my mother or Ben's family were going through at the time, but after almost two years of uncertainty, it had to be a trying and stressful period for them.

As time went on, springtime conditions worsened in the camp. Men were cooking outside on small cooking stoves, and smoke pervaded the camp. It was still heavily infested with fleas, lice, and bed bugs so bad that more and more of the men elected to sleep outside. Eventually, huge white tents were set up to handle the growing number of prisoners, and some men slept in trenches (Figure 9-10). When it rained, the men in the tents could not stay dry. Mud and rivulets of water flowed on the bare ground. In addition to tents, there were crudely constructed lean-tos, and four hundred men were in barracks designed to hold one hundred. Twice a day, there were appells, but counting prisoners seemed useless, as the Germans had no idea what an accurate headcount was. A long trench was the common open-air latrine, and cases of dysentery were rampant.

Figure 9-10: Stalag VIIA in Moosburg (photo courtesy of Clark Special Collections Branch of the USAF Academy Library)

According to Bub Clark, "Most of our men were utterly miserable. They were hungry, cold, discouraged, crowded, and being eaten up by the bugs in our dirty blocks… Each man had a mass of flea bites on his chest."

The German rations were much worse than in Stalag Luft III, and only a few of the normal Red Cross parcels were periodically distributed.

The stoves in the barracks were inadequate to cook the very meager quantity of food, so the Kriegies made little stoves out of tin cans and cooked food outside on them. Shown in Figure 9-11, the stoves consisted of a blower spun by a crank that was connected to the stove. A small pot could then be put on the stove to heat up water or cook food.

Trying to feed 130,000 hungry men became impossible. Cabbage was in abundance, so it was turned into sauerkraut dumped into large wooden barrels that sat for long periods of time and made everyone sick. My father hated sauerkraut for the rest of his life, and it is no wonder that he would not eat it.

Fortunately, with the intervention of the International Red Cross, the YMCA, and the Allied forces' hierarchy, agreements were made on how food and medical supplies could be transported to the large camps where prisoners had been consolidated. The increased influx of Red Cross parcels and YMCA supplies made a huge difference to the POWs, who were severely malnourished and in need of medical care.

The poor infrastructure of roads and railways in Germany due to bombing as well as the lack of fuel hindered progress. Luckily, the International Red Cross was able to obtain fuel and used marked vehicles, trains, and even ships to deliver goods. This was related in my father's journal when he wrote of vehicles transporting parcels from the Red Cross headquarters in Switzerland. It had been a full month since the long winter marches, and as of March 6, 1945, the convoys were permitted by the Germans to travel safely to the prison camps.

Unfortunately, as my father noted, some of the contents of the Red Cross parcels were stolen by the Germans before they were delivered to the prisoners. Stolen from his parcel were a pipe, a bar of Swan soap, four packs of Camel cigarettes, a plug of chewing tobacco, a pair of dividers, a can of coffee, a box of raisins, and a chocolate bar—as he mentioned, "a serious loss." After that, he was out of Red Cross parcels and back on rations.

On March 9, due to the new convoys, it was officially announced that two weeks of parcels had arrived. Ben and his combine received one American parcel for four men. After that, they only received British Red Cross parcels, which were basically the same as the American ones except they had tea instead of coffee. Receiving the packages was a huge morale booster for the prisoners, as they not only provided sustenance but also necessary personal items.

Figure 9-11: Americans cooking on a makeshift blower stove made from KLIM (powdered milk) cans (photo courtesy of Moosburg Online)

Swedish lawyer Henry Söderberg, as a representative of the International YMCA during World War II, was one of seven foreigners allowed within Nazi Germany to visit POW camps. After he learned about the evacuation of Stalag Luft III, he traveled through war-torn Germany and struggled to find gasoline for his vehicle. He finally reached Moosburg and Stalag VIIA on April 16. As described in Arthur Durand's excellent book, *Stalag Luft III: The Secret Story*, Soderberg was shocked at the condition of the prisoners, stating, "Thousands and thousands of them live in barns and stables and bombed-out houses and in the small forest around Moosburg. There are new streams of prisoners marching in every day." He then worked hard to distribute food, medicine, and other necessities for the prisoners there.

Due to the overcrowded conditions and other problems at Stalag VIIA, the senior American officers were challenged to keep good mental and physical health among their POWs. SAO Colonel Paul R. "Pop" Goode and his staff attempted to keep morale high by enforcing strict discipline among the American POWs. He had them thoroughly organized, with each man doing daily tasks. Saturday inspections were held, in which each officer and man was required to show up freshly shaved and bathed with his hair properly cut. Personal cleanliness was reflected in the well-kept but crowded barracks and tents, which were closely policed under Goode's direction. This discipline and the additional Red Cross assistance helped to keep the prisoners healthy and mentally alert.

On March 22, Ben had his first hot shower since entering the camp and received a pair of socks and two handkerchiefs from the Red Cross. With the dismal conditions at Stalag VIIA, these were real perks and ones that were much appreciated. There were a few other perks, such as a celebration on Easter, April 1, when there was a "big bash" with lots of desserts!

With an increasing number of prisoners at Stalag VIIA, my father moved yet again on April 9. That time, it was to another crowded area at the back edge of the camp. Unfortunately, he was housed in a tent but was able to see many old friends.

USAAF standard issue for pilots were the brown leather bomber flight jackets. They were valued possessions, and I often wondered what had happened to Daddy Ben's. Was it confiscated, lost, or destroyed when he was shot down, captured, and imprisoned? After reading his journal, I found out! On April 15, 1945, shortly after his twenty-sixth birthday, he wrote: "Traded my leather jacket, which was in pretty bad shape, for a field jacket, M1943." To my astonishment, he'd kept his leather jacket from the time he was shot down, while he was a prisoner at Stalag Luft III, and during the long march. I'm sure that it was hard to give it up after all they had been through together. However, my father was a practical man, and this act showed the state of his desperation.

Chapter Ten
Liberation

On April 19, Ben was on the move again. He shifted to a compound across the street from where he had been. Fortunately, he was in a barrack again (Block 14) with lots of room to walk around. In an unusual move by the Germans, the gates between the compounds were opened. This gave the prisoners freedom to walk about and meet or become reacquainted with other prisoners. Prisoners made new friends from different nations and different services. In some cases, there were reunions of friends from long ago.

My father finished his version of the journey:

We are here as I write this and have been shuffled about all over the place. I've lived in stables, tents, and flea-ridden barracks and am now in a fairly clean and orderly barrack.

There is a strong possibility that we shall be evacuated from here, but the weather is warm now, and conditions are different. We've grown used to the gypsy life, and anyway—the war is almost over! April 22, 1945

As Patton's Third Army closed in, Hitler wanted to move the British and American flyers deeper into Bavaria with another march. However, an agreement had just been made between Berlin, London, and Washington. The Germans said that Allied POWs would no longer be moved if the Allied governments would agree that, when liberated, those prisoners would not be thrown back into fighting in the war. The United States, Great Britain, and Soviet governments accepted this proposal, and it went into effect at midnight on April 22.

Fortunately, the Allied senior officers in Stalag VIIA learned about the agreement by listening to BBC and Radio Luxembourg. The next morning, when the Germans instructed the prisoners to line up at the gate for another evacuation, the Allied POWs refused to move. The Germans in the camp checked with their supervisors, and indeed, the Allied POWs did not have to be marched again. As my father wrote in his journal on April 24: "Announced that we would not evacuate. Have been sweating it out for several days."

Through the clandestine radios still available, the Allied prisoners could track the advances of the American troops. There was hope for liberation in a matter of days. However, the POWs still did not know what would become of them if the Allied forces did approach the prison camp. Would they be shot by the Germans, would they be held hostage, or would they be freed?

The camp became increasingly unruly and could be a dangerous place if a man wandered into a foreign area when barbed wire had been pushed down. Desperation was constant and dictated behavior. Much trading went on for food between the Allied POWs bargaining with other nations' prisoners.

By the end of April, word quickly spread that the war was almost over with the takeover of Nuremberg on April 20 by Patton's troops and the bombing by Allied aircraft in nearby Munich. Allied fighter planes flew over the camp and waggled their wings to a cheering crowd. Every day, more planes flew over, some on bombing missions. Ben saw B-17s in the air and was thrilled by the sight. On April 25, he exclaimed in his journal: "It seems only days until we'll be liberated!"

Unbeknownst to the POWs, SAO Colonel Goode and SBO Captain Kellet left Stalag VIIA and went to the headquarters of Combat Command A, 47th Tank Battalion, in the early morning hours of April 29, 1945. They carried a proposal from the Germans that the area around the prison camp in Moosburg be declared a neutral zone until the Allied and German governments could decide what to do with the prisoners. This was summarily turned down by the American commanding officer because it would give the Germans more time to retreat and possibly move the POWs. The German proposal would also affect the American troops' plan to take over some of the bridges in Moosburg, as they would be in the neutral zone. In return, the Americans asked for an unconditional surrender by the Germans.

If the Germans did not respond to this request by 9 a.m., the American division commanding officer, Maj. Gen. Albert C. Smith, would order commanding officer Brig. Gen. C.H. Karlstad to attack Moosburg in the morning. Prisoners were instructed to stay in their barracks because bullets would be flying, so everyone took whatever cover they could find.

Resistance from the Germans was strong but short lived. The Germans had no tanks and were armed only with small arms, machine guns, mortars, and Panzerfausts (anti-tank guns). They showed their strongest force about a mile west of Moosburg, when they unsuccessfully tried to stop the Army troops from crossing the Amper river. The battle-hardened Americans fought their way through and continued to Moosburg. It was not long before the POWs could hear the cannons and guns in battle in the distance. SS troops valiantly but unsuccessfully began firing their small arms from the cheese factory and in church steeples in the vicinity of the camp. Some of the small arms fire from this attack whizzed over the prisoners, who had hunkered down.

The Germans surrendered around 10:30 a.m. as the Americans reached the edge of Moosburg and continued to roll through. Then, the 14th Armored Infantry Division of Patton's Third Army barreled over the hills approaching Stalag VIIA.

As the firing stopped, the Stalag VIIA prisoners cautiously came out of hiding, and some brave men stood on the roofs of their barracks. From there, they could see the column of American Sherman tanks on their way to the prison camp.

Three American Jeeps with machine guns entered the prison camp and demanded that the German guards and commandant surrender and hand over their weapons. Moments later, a battle-scarred medium tank joined them at the main gate and rolled right through the barbed wire fence to shouts of joy from the prisoners. The POWs went wild! They were climbing all over the tanks and celebrating, as shown in Figure 10-1. They cheered and cheered—they were free!

Figure 10-1: Liberated prisoners climbing onto tank from Patton's Third Army that entered Stalag VIIA on day of liberation, April 29, 1945 (photo courtesy of Clark Special Collections Branch of the USAF Academy Library)

One of the prisoners reflected on that day years later. "You could hear the tanks coming before dawn," John Lee Kirkpatrick recalled. "Then, you could see the outline of the tanks. The GIs with goggles on their faces, sandbags on the fronts of those tanks, and little American flags stuck in the sandbags. And you could see that ... that big white star on those tanks." He added, "That was the most moving experience."

Now, freedom came rolling through the front gate of Stalag VIIA on the treads of American tanks. "We had been waiting two years, and we thought it might be five," Kirkpatrick said. "But now, it was over. The tankers began throwing out rations," he said. "They gave us every bit of food they had."

The most emotional moment came a few minutes later. My father chronicled the momentous event, stating that at 12:40 p.m., the Nazi flag shredded by bullets was taken down from a nearby church steeple, and the American flag was raised after about three hours of confrontation between the German and American troops in the vicinity.

At 1 p.m., at the main gate of the camp, an American POW, First Lieutenant Martin Allain, ripped the Nazi flag off the flagpole. Then, using an American flag entrusted to him by Lieutenant Colonel Clark, he hung the Stars and Stripes on the flagpole. What a poignant and exciting time that was for the prisoners, who had not only cheers but tears flowing.

I saw that Nazi flag at the Mighty 8th Air Force Museum in Savannah, Georgia, and it was signed by many of the POWs. It was an emotional sight for me, as I imagined that my father had been overjoyed to see it come down. He exclaimed in his journal: "I'm a free man again—after one year and nine months as a POW." After eighteen months in Stalag Luft III, the long winter march, and three months in Stalag VIIA, my father was finally free.

Bob Doolan gave a different perspective on the raising of the American flag: "Hard to keep from crying. Most wonderful sight in my life. There was no way to describe our feelings. It was not so much a celebration. It was [as] if a heavy weight had been lifted not only for us but off the whole world. Men wandered around in groups, or alone, some with a vacant look, as if they could not comprehend the whole thing. I can't remember ever going to bed."

In 2022, I was able to visit Moosburg and the site where Stalag VIIA had been located. There was a commemorative display where the gate for the prison used to be. All I could think about when I stood in that spot was that Daddy Ben had marched through that gate after arriving there from Stalag Luft III and had ultimately been liberated through it. It was a meaningful and memorable experience for me, and I was glad that one of the prison barracks and two of the guard barracks were still standing.

At liberation, the Red Cross came into the camp and tended to the prisoners, giving them food. Two days later, on Tuesday, May 1, General George S. Patton arrived in his Jeep with his four stars plastered everywhere, even on his ivory-handled pistols. The prisoners cheered him on and knew that their liberation was a result of his actions, and their time as POWs was finally over. Army field kitchens were set up to feed the hungry men. My father was ecstatic, writing, "Had first white bread—with orange marmalade!" He exclaimed later in life that white bread tasted like cake to him.

Unfortunately, liberation and freedom were not immediate. Due to the sheer number of POWs at Stalag VIIA, it was not possible to quickly evacuate the camp. There were 130,000 Allied prisoners, fifty thousand of which were Americans. The war was still going on around the prison camp, and there were groups of SS nearby. The SAOs did not want the former prisoners to get shot trying to leave the camp, so the former POWs were ordered to stay. Many violated that order and went out exploring.

The American POWS were told not to escape or leave the camp; however, many did. Some prisoners from other countries left and scavenged food from nearby residents through bartering and downright robbery. Then, those men would return to Stalag VIIA and tell their tales.

While many men grew impatient and hitchhiked their way to Paris, my father stayed at Moosburg and was able to take a walk in the woods around the camp on May 2. Those who left the camp grabbed any vehicle they could, from bikes to horse-drawn wagons to beautiful German cars, to make their way to Paris.

Days were full of frustration and disappointment, especially after such elation on April 29. The German guards had been taken to Allied prisoner of war camps, yet the camp in Moosburg still had to function. It took a while for the U.S. Army and the Red Cross to provide the necessary support for the newly liberated prisoners. The Allied forces needed time to coordinate an orderly evacuation, and the former American POWs were organized under their already appointed leaders.

As the time for the men to head home drew closer, they were called RAMPS (recovered Allied military personnel), and my father was glad to fill out the registration form, which I have, to expedite his evacuation. The process had begun to fly out the men to Camp Lucky Strike in northern France at Le Havre where troop ships could come into harbor. Time as a POW and rank would determine the order of returned men.

On May 7, 1945, the Supreme Headquarters Allied Expeditionary Force commander ordered that all American and British ex-POWs be evacuated from the liberated areas in the shortest possible time, regardless of any limiting circumstances. Combined Air Transport Operations Room, the agency that controlled air transportation, was given the movement of the POWs as its top priority.

On May 8, 1945, my father was flown by C-47 aircraft to Liège, Belgium, and then transferred by truck to the reception camp in Namur, Belgium. For most of the freed American prisoners, it took approximately ten days to get out of the awful conditions at Stalag VIIA. Most were taken by trucks to Straubing, France, where shot-up German planes still littered the grounds. Then, they were flown in a C-47 to Camp Lucky Strike in Le Havre, France. A picture of some of the former POWs leaving on a C-47 is shown in Figure 10-2.

Figure 10-2: Allied RAMPs board C-47 aircraft in Germany bound for France, one of the stages of their evacuation to freedom, April through May 1945 (USAAF photo)

Former POWs were given a physical when they arrived, along with backpay. POWs, so long half-starved, were only allowed to drink eggnog through the day until their stomachs could handle solid food. They were supplied with food, clothing, bedding, and toilet articles at the reception camps. They stood in long lines waiting for everything, which was very tiresome to them, and then transported as soon as possible to the embarkation staging areas for a trip back to the United States.

At the reception camp at Namur, Belgium, Ben was processed in and received clothing and other goods. He stated that he was quartered "in fine style," and I have his mess ticket from there. He enjoyed a welcoming meal full of nutritious, good food. At his physical exam there, he weighed in at 146 pounds. He was thin for someone with a five-foot, eight-inch frame but was in reasonably good health. After a day of relaxation, he went to Camp Lucky Strike by train.

In France, the Americans had set up camps for the RAMPs. As they could not be numbered or named like other military bases or prison camps, they were named after cigarette brands to avoid confusion, as shown in Figure 10-3. The main camp about forty-five miles from the Embarkation Port of Le Havre was Camp Lucky Strike near St. Valery-sur-Somme. It was the primary staging area for recovered American POWs. There were nine other such camps—eight in France and one in Belgium.

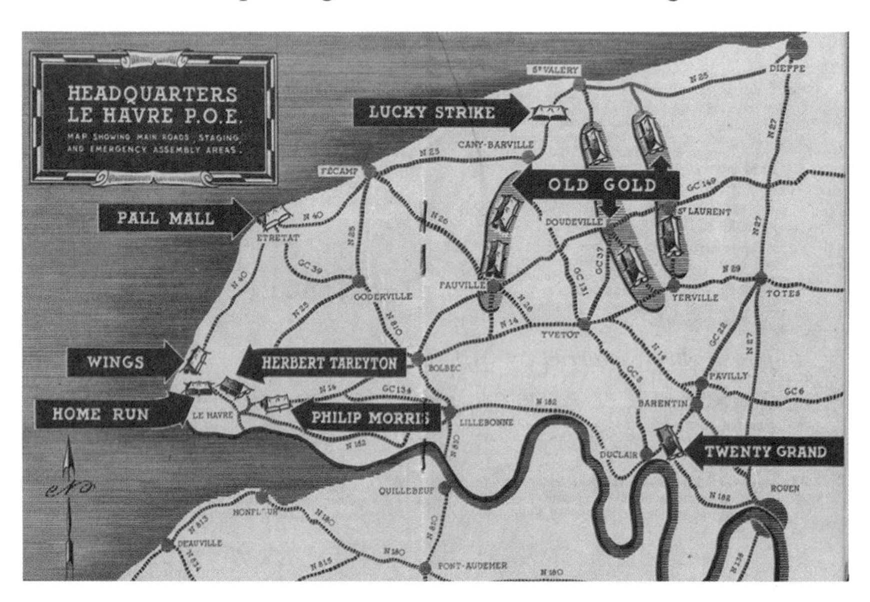

Figure 10-3: Army Transportation Corps booklet, You're Staging for the States (Photo courtesy of World War II Museum, Gift in Memory of Charles S.M. Coddington, 2011.373.004)

As the war ended in Europe, 73,000 troops were sent to Camp Lucky Strike for processing and return to the U.S. It was the largest camp and the chief assembly point. As described in an article on Camp Lucky Strike by Kim Guise, WWII Museum curator, it was, "Seventh Heaven and complete chaos." Processing consisted of showers, delousing, and issuance of new uniforms and equipment. The troops underwent psychiatric and medical exams (with X-rays, blood draws, urine samples, and other tests) and were given necessary inoculations. Many of the men required hospitalization due to respiratory infections or malnourishment. RAMPs in need of convalescence (some arrived via hospital train) were assigned to Block C.

By May 1945, Camp Lucky Strike had become very well organized. Reception into the RAMP camp was in a designated area with a capacity of about two thousand beds. The men were kept there for a day, after which they were transferred to the processing areas, with a capacity of four thousand beds, for a minimum of one and a half to two days. After processing, they proceeded to the "pending shipment" area for a minimum of one day (though in actual experience, it was three and a half to ten days).

The biggest thing on most of the former prisoners' minds was food. Especially after the meager rations at Stalag VIIA, they were malnourished. Food stations were set up at Camp Lucky Strike—staffed by German POWs, interestingly enough. Food intake was regulated for the former Allied POWs so that their digestive systems could adapt to a healthy diet. The former prisoners were anxious to eat as much as they could but were cautioned to only eat limited amounts, or they would get sick. Lectures were given and brochures were provided to educate the men on how to eat properly without getting diarrhea or other gastrointestinal problems.

However, the first groups of POWs were met by the American Red Cross people, who provided hot coffee, doughnuts, peanuts, and a wool blanket. Free doughnuts were handed out—one per person. Not surprisingly, the recovered Kriegies were masters at bargaining and bartering and accumulated more than one ticket apiece! Unfortunately, gorging on doughnuts and peanuts led to many being sent to the hospital. A bland and soft diet was required for at least forty-eight hours until their systems could accommodate regular Army rations.

My father arrived in Camp Lucky Strike on May 11 and was quartered in a tent. I have a mess ticket and an exchange ticket from my father's time there. The mess ticket stated, "This certificate entitles bearer to initial and weekly gratuitous issues of exchange supplies during period stationed at Camp RAMP. This slip must be kept by RAMP and presented whether for initial or subsequent issue." The U.S. Army Exchange ticket said, "The bearer is authorized to purchase not more than two gifts or one perfume at the RAMP Gift Shop." I guess that French perfume was a hot commodity!

Another part of the processing at Camp Lucky Strike involved documentation with a number of forms and applications to be filled out. Then, there were long and onerous debriefings. The men were asked to describe their missions, capture, and time in the prison camps. Some of the information was used to locate men still MIA, and some information was needed to prosecute those who had participated in war crimes. Identification tags were distributed and were required to be transported back home.

At Camp Lucky Strike, those awaiting transportation back to the United States were in a bustling tent city of over seventy thousand troops. The tents were lined up alphabetically with four letters or so per tent region. Each region had its own infrastructure with churches, post exchange shops, barbers, and auditoriums. Entertainment and movies were provided to try to keep the anxious troops busy, but restlessness and impatience were rampant.

Time was also spent in the reuniting of friends and colleagues and in enjoying the good coffee provided by the Red Cross. Knowing Daddy Ben's personality, I am sure that he was relieved and happy to be out of prison camp after being held for almost twenty-two months. He probably enjoyed the good food and entertainment and cheerfully socialized with fellow airmen whom he had not seen for a long time. I hope he reconnected with the six non-officer crewmembers who were shot down with him and sent to different prison camps. He did find several of his friends there and listed them in his journal: "Fogarty, Pearson, O'Connell, and Chandler." According to him, there were "lots of old friends around."

Men could send one cable home, and others sent postcards. A note, transcribed and typed out by my aunt, that was possibly written by Daddy Ben while he was in the RAMP camp expressed his elation:

A free man again! I guess I never realized before just how much freedom means to a man. I found that in addition to the big things—a wife, a home, a family—there are a lot of little things that count.

I never realized what a privilege it was to walk in a straight line, for instance, or to lie on soft, green grass in the sunlight. These little things I missed for the first time because I had never been deprived of them before.

The United States War Department dictated that the first men to be shipped overseas were POWs who had been held more than sixty days. Men who were to be shipped were notified seventy-two hours prior to departure and trucked to staging areas at the port. Before leaving, their personnel orders were compiled and checked, medical checks were conducted, and baggage was inspected. Each person was allowed only twenty-five pounds. Once again, Ben's Wartime Log, his "PENNY" comic strips, and the letter from my mother survived the cut.

Before he left Camp Lucky Strike, my father sent a cable or a letter to my mother:

You should know by now that I am a free man again and will soon be on my way home to see you! Gee, it seems too good to be true, doesn't it, baby? We've waited so long and prayed so hard. You are such an angel, and I love you so completely, and I can't tell you how very proud I am of you. You have taken it all so well, honey—and I know how hard it must have been. I love you, sweet, totally and completely, forever.

It was April 29 that we here at Moosburg, about twenty miles north of Munich, were liberated. Within a few days, I'll be on my way to you, and each mile I now travel will bring us closer together. Though I don't know exactly when I'll see you, it should be this month or the next. Each day will seem like an eternity until I hold you in my arms again. We'll have so much to tell each other.

I am still in good health, angel—we can be grateful for that. There were times when we weren't too well fed and times on our February march when I wondered if everybody would make it—but everybody did make it in my camp. Well, that's all over now—we are no longer prisoners. We have more than we can eat now—even white American bread! I know now what freedom means and can certainly appreciate it all the more.

I'll write you every chance I have, darling, and will certainly let you know if you can write me. I'll call you if I should ever get a chance and will of course let you know as soon as I find out when I'll be home.

Give my regards to all the folks and tell them that I'm okay. So be sweet, angel, take it easy, and I'll see you soon! I love you!

On Sunday, May 20, Daddy Ben was moved to the "D" area of Camp Lucky Strike. On May 23, he was transported by truck to Le Havre (and said there was not much there!), which was the "pending shipment" area. The former residents of the German prison camps, as well as other military personnel, traveled in overloaded ships for over a week's journey to the East Coast of the United States. What a welcome sight that would be! My father's luggage tag for the trip home showed his officer number and group number, GMO 1029, 10-74.

From the Le Havre port, there were up to six thousand people departing a day, with 207,759 people shipped out at its peak in the month of June 1945. In total, three million American troops entered or departed the Le Havre port, and more than 73,000 RAMPs were processed through Camp Lucky Strike.

On Friday, June 1, Daddy Ben boarded the ship *SS Marine Robin*, which carried him from France to New York. The *Marine Robin* was a variant design of the basic C4 ship, originally designed for American-Hawaiian Lines and taken over by the United States Marine Corps.

On June 2, 1945, freed POWs were addressed on the *SS Marine Robin* radio by General Eisenhower before departure. He explained that he would like for them to have "first class" accommodations, but there were so many freed POWs, the ship could not take them all without double loading, meaning two men sharing a bunk alternately in twenty-four-hour periods. The men were given a choice to travel without double loading, leaving many behind, or to endure the hardship. They voted to double bunk and leave no one behind, departing the dock at Le Havre in the evening for the return home.

I was fortunate to find my father's mess ticket from the *SS Marine Robin*. Group number two, table seven, seat twelve was printed on it. Breakfast was at 0715, lunch was at 1315, and dinner was at 0615. His ticket stated, "You are allotted one-half hour for eating. As soon as you have finished, leave the table and dining hall in order that the next group may be served promptly."

The ship *Marine Robin* transited to New York unescorted and arrived June 8, 1945, for a brief period of quarantine before disembarking in Manhattan to board trains for Camp Shanks. Camp Shanks had previously served as a staging area for troops departing the New York Port of Embarkation for overseas service during WWII. In 1945, it was used as a processing center for those troops returning from the war.

I can't imagine my father's feelings upon entering the New York harbor and seeing the Statue of Liberty. I know he entered through Camp Shanks because my mother's scrapbook has a "coupon" of his place in line to make a telephone call. He was in Group 1029-10B, and the coupon gave him permission to make a phone call on June 11, 1945.

Who did my father contact for that first phone call? I would like to think that it was my mother. Whoever he called was surely ecstatic to hear his voice! What did he say? What did they talk about? I'm sure there was a long line behind him, so the conversation was probably short. He was happy and excited, and I'm certain that everyone cried tears of joy. There was so much news to share from the events of the last two years, but that phone call let everyone know that all was well.

My father was then flown to Dallas, Texas (Love Field), to be reunited with my mother and his family. He arrived there on June 17, 1945. There was a cheering throng when the airplane carrying all the WWII veterans landed! There are many pictures of veterans greeting their families and kissing their brides, but unfortunately, I have no photos of my parents from that day. So, I will just have to create those vignettes in my mind.

During WWII, 16.5 million people valiantly served in the military, and 73 percent of those served overseas. Among those who served, 130,201 were POWs; 124,079 were Army and Air Forces personnel, and 6,122 were Navy and Marine Corps personnel.

The POWs endured varying conditions and withstood capture and imprisonment. They learned from their experiences, and most felt it made them stronger. They forged lifelong friendships, and many participated in POW reunions over the years. In particular, South Compound POW reunions were organized by Bub Clark for several years and had interesting speakers like German prison guards, German censors, and members of the Army who liberated Moosburg. These gatherings were characterized by many happy greetings and catching up with forever friends.

I don't have any record that my father participated, although his copilot Dave Pollak was very active in the reunions and helped to organize them. However, Daddy Ben was a card-carrying member of the Texas Kriegie Horyo Club and the Texas Prisoner of War Club!

I know there were long-lasting effects of post-traumatic stress disorder—nightmares and flashbacks and terrible memories—for many POWs. I don't know whether my father experienced any of these things, but he never talked about the war or about being a POW. I believe these topics were compartmentalized in his mind and he moved on, preferring to put them behind him.

Doing research for this book has really opened my eyes and "filled in the gaps" for what Daddy Ben went through during his time in Stalag Luft III. I learned so much about my father and the type of courageous man he was. I believe he used his creative talents to be resourceful and to improvise new solutions with what he had available. I also believe that he used his time there to the best of his ability—drawing the "PENNY" cartoons, painting and writing in his Wartime Log, drawing pinups in other men's logbooks, carving wood, cooking, and probably participating in sports, theater, and musical performances. Above all, he was a team player who was part of the camaraderie that aided in survival.

When I was in the astronaut corps, I had this same sense of camaraderie that my father shared with his fellow prisoners. The circumstances were entirely different, but there are some similarities. People in the NASA astronaut corps were from a mix of backgrounds—military officers, civil service, pilots, doctors, scientists, engineers, and many others. We also had astronauts from other countries who were training alongside us either as mission specialists or payload specialists. All had one thing in common: We were risking our lives and facing danger whenever we flew into space. This common bond unified us in a way I have not experienced with any other group of which I have been a part. The danger and risk were part of our jobs, and we accepted this in service to our respective countries. We didn't talk about what could happen to us in flight, but we all knew.

In my father's case, he also was a part of a diverse group of men who faced danger and took risks while waging a war. After some situations of extreme danger, being shot at, and having dealt with a crippled airplane and/or crew, they were taken prisoner by the Germans. Despite their different backgrounds and nationalities, their time as POWs was a unique experience that would forever bond them and ignite solidarity wherever they went.

Chapter Eleven
You're in the Air Force Now

I wonder what was going through my father's mind as he was flown home to Texas from New York. At the same time, while my mother was waiting for his arrival, I wonder what she was thinking. They were different people in the summer of 1945 than they had been two years before. They had not seen each other since that sunny day near Orlando when they'd kissed goodbye, not knowing whether they would ever see each other again.

Each time I went into the dangerous realm of space, my mother went through similar emotions of apprehension and joy when she learned that our Space Shuttle had safely landed. Until she heard the "wheels stop" announcement, she was scared and praying. She never let me know how frightened she was until it was all over.

My mother was nineteen years old when my father left for Europe, and during the ensuing two years, she had experienced loss, uncertainty, heartache, and longing. Most of the time when Daddy Ben was at war, she had worked or gone to college. She had grown from a teenage bride into a mature officer's wife. Of course, she was anxious to see her husband return from war, but I'm sure she had many concerns.

My father was twenty-four years old when he started the ferry of his YB-40 across the North Atlantic. Since that time, he had watched in horror as his colleagues and their aircraft had perished during his seven combat missions and additional aborted missions. Even when he hadn't flown in a combat mission, he'd felt loss and sadness when there was an empty bed at night after a friend did not return from war. It had also been a time of learning and exhilaration, as he'd experienced living in another country and discovered so many new things about life and loss.

During his trek home from New York, he was full of excitement and wonder as well as relief. When he arrived to hugs and kisses, he was thinner and more mature. He had witnessed war firsthand and had much to share. He and my mother were happy and full of hope, not knowing what the future would bring but glad that the past two years were behind them. All were thankful that Ben was alive and that the war was over for him.

The Germans surrendered to the Allied powers on May 8, 1945, ending WWII in Europe. Many of the aircrew and ground crews who fought in Europe were transferred to the war in the Asia-Pacific. Some were readied to be shipped to the Pacific but were never deployed there. The United States dropped their first atomic bombs on the Japanese cities of Hiroshima on August 6 and Nagasaki on August 9. Fearing more atomic bombings, the Japanese announced its intention to surrender on August 14 and formally surrendered in September 1945, ending WWII in the Asia-Pacific.

Each American POW who returned from Europe was given sixty days of paid leave. Just as my mother welcomed Ben back into her life in 1945, I have welcomed a new part of him into my life in 2022. I was just as eager to know him better, as my mom was eager to reacquaint with him so many years ago.

After indulging in conversation, good southern food, and family reunions, my mother and father set about continuing their young marriage. They were much different now, and it would take years to reconnect—if they ever could. So, they took mini vacations during his leave time. For example, I have a receipt in my mother's scrapbook of a visit to San Antonio from June 16 through June 18 at the White Plaza Hotel. Each night cost them $3.50!

While Ben was acclimating to a normal life, my mother was acclimating to having a war veteran back in her life. All these adjustments were taxing on these two young people, and their stories reflect many more like them after the war. They had both changed so much, and they were starting their relationship all over again.

Ben was on leave June and July of 1945. However, he was anxious to get into the cockpit again, as he had not flown as a pilot since that fateful day on July 28, 1943. Almost two years later, on July 20, 1945, he climbed into the cockpit of a B-25 airplane as a copilot, reveling in the sound of those engines revving up and his plane once again rising into the sky. In his flight records, the flight was listed as "casual," and I'm sure that he thought so as well since no Germans were shooting at him, and the flak was absent! He flew for almost five hours and logged ten landings.

There were other casual flights in B-25 and C-47 aircraft, with many hours and many landings made by Ben out of Love Field in July 1945. The time was voluntary because that's what pilots do—they love to fly. I can certainly relate to that, as it is always a thrill for me to be at the controls of an airplane (or jet) and to experience the joys of flight. It is noteworthy that he flew a C-47 aircraft—the very type of plane that rescued him from the bowels of Germany and transported him to freedom in Camp Lucky Strike.

My father was assigned to San Antonio for the months of August and September 1945, as he decided to stay among the ranks of the USAAF. He rose to the rank of captain, and a picture of him post-war after he achieved those captain's silver bars is shown in Figure 11-1. He remained on the pilot rolls, though he did not log any flying time while in San Antonio.

Figure 11-1: Captain Ben Smotherman after returning home from WWII (USAF photo)

With Ben's experience in the war and as a USAAF officer, he gathered and disseminated knowledge about missile technology. Now that the war was over, bombing from aircraft piloted by people became a smaller part of the strategy for waging war from the air. More emphasis was then given to pilotless missiles and higher technology bombs that could be dropped either remotely or from a high-altitude aircraft. Ben would not fly in combat any longer but would help those who did in the future.

In January 1946, after my father had more training and proficiency flying, my parents had a permanent change of station to Chanute Field in Illinois. It would be a new experience for them because they had never lived together as a couple for any length of time or had their own home.

As mentioned previously, while my father had been in the war, my mother had continued her education at Southern Methodist University, accumulating more undergraduate hours. I know that she valued education and wanted Ben to earn his bachelor's degree. Although at Chanute working full time as a USAAF officer and flying as a pilot, he also enrolled at the University of Illinois under the Servicemen's Readjustment Act, studying civil engineering. My mother continued her undergraduate studies by majoring in industrial psychology at the University of Illinois, earning her Bachelor of Arts in February 1951. During their time in Illinois, my parents bought their first home and moved from the Chanute base housing that was provided.

The War Department, led by Army General Dwight Eisenhower, supported the effort to create a separate air component out of the AAF. On July 26, 1947, President Harry S. Truman signed the National Security Act of 1947, which established the Department of the Air Force, but it was not until September 18, 1947, that the United States Air Force (USAF) was officially formed.

Among all the activities that resulted from forming a new branch of the military, my father traded in his Army olive green uniforms for the Air Force blue uniforms. These looked very much like the Army uniforms in style but were different in color. The ranks, rank insignia, wings, and other types of recognition were carried over to the USAF from the USAAF.

Other changes in the Air Force during my father's time at Chanute evolved from a different strategy of warfare and air power. With the advent of the atomic bomb and the demise of strategic bombing by aircraft, technology had advanced to the point that warfare by missiles was advancing rapidly. As a new way of attack, various ground-based missiles were being developed. My father became involved in this type of warfare and the development of these missiles.

In this timeframe, President Truman made the decision to bring German scientists and engineers in aeronautics, electronics, medicine, materials science, and rocketry into the United States. There was a fear that if these specialists, some of them Nazis, went to the Soviet Union, their expertise could be used against the U.S. Called *Operation Paperclip*, it was a highly secret mission. Included in this group were Wernher von Braun and 127 of his rocket scientists, who were instrumental in developing the V-1 and V-2 rockets for Germany, which were used in WWII to bomb London and other locations in Europe. The Germans were kept at Fort Bliss, Texas, in late 1945 until their permanent location was determined. The U.S. Army decided that the rocket ordnance for the Department of Defense would be kept at Redstone Arsenal in Huntsville, Alabama. Therefore, on April 1, 1950, von Braun and his team and their families moved to Huntsville.

Operation Paperclip and the acquisition of this tremendous human talent in rocketry may not seem to have direct applicability to the story of my father and his career. However, my father was involved with missiles—our partners in rocket propulsion—and the work in Huntsville was a factor in the development of missile and rocket technology for our country. Furthermore, von Braun and his team laid the foundation for our space program and were fundamental to the culture and environment of Huntsville, Alabama, where I was raised.

While at Chanute, Ben continued his flying, technical training, and university studies. In September 1950, my parents left the cold climate of Illinois for the muggy southern town of Biloxi, Mississippi, to Keesler AFB for more training. At Keesler, Ben also attended the Airborne Electronic Officers School. He learned about radar technology, as it was necessary to track the trajectories of missiles that had launched. He also logged several ground-controlled approaches in the T-7 aircraft, helping to train the radar specialists and ground controllers at Keesler. My parents stayed at Keesler almost a year—until June 1951.

As Ben was a pioneer in the field of guided missiles, in July 1951, he was sent to participate in the development of the Matador tactical missile at the newly activated Patrick Air Force Base (PAFB), south of the town of Cape Canaveral, Florida. The development of Matador "pilotless bombers" had begun shortly before the end of WWII and was accelerated into production in 1951 due to the Korean War. The Matador was a surface-to-surface missile capable of carrying a conventional or nuclear warhead. It was part of the U.S. military's search for methods of delivering heavy weapons payloads long distances without risking human crews.

Figure 11-2: Matador surface-to-surface cruise missile, or "pilotless bomber" (photo courtesy of 45th Space Wing History Office, Patrick Air Force Base, Florida)

A picture of the launch of a Matador at Cape Canaveral Air Force Station is shown in Figure 11-2. I also have some pictures of Ben testing the Matador while he was in Florida (Figure 11-3), but my favorite picture of him at PAFB is the one of him with the missile he named after my mother (Figure 11-4). It was called "Dolly Jo, Ben's Bride, Ballerina of the Atmosphere." Ben continued to pursue his love of flying at PAFB and piloted C-47s, B-17Gs, B-25s, C-45s, and B-29s as part of his job. He was fortunate that he was still able to continue flying and to have a successful career as a pioneer in the burgeoning missile technology during that time.

Figure 11-3: Ben Smotherman at the controls of a Matador launch at Cape Canaveral, Florida (USAF photo)

Figure 11-4: Ben Smotherman with a Matador missile named after his wife, Dolly Jo, at Patrick Air Force Base, Florida (USAF photo)

My mother and father were delighted to return to Florida, and after selling their home in Illinois, they bought a cement block house on the beach on what is now Highway A1A. There were not many houses around them, and they had a view of the ocean! A picture of the house is shown in Figure 11-5, and it still stands today. My parents thoroughly enjoyed living on the beach, and for the rest of their lives, each of them longed to live in Florida again near the ocean.

Figure 11-5: Jan's first house on Cocoa Beach, Florida

My mother fondly remembered the time that she lived in the beach house in Florida and said it was one of the happiest periods in her marriage to Daddy Ben. My father had become a major in the USAF on June 1, 1952, and their social life was full of parties, dinners, and other occasions befitting an Air Force officer. I am sure they visited the PAFB Officers' Club, which was on the beach. Coincidentally, that same Officers' Club was the location of my prelaunch reception for guests who came to Florida to witness my first Space Shuttle launch in 1992, some forty years after my parents lived there.

My father continued his drawing and painting, using not only watercolors and acrylics but also oils. Several of his oil paintings adorned their house in Florida. Because my mother was a dancer and collected ballerina figurines, Daddy Ben painted a series of ballerina pictures for her. Today, I have one of them hanging in my house. As you may recall, during his time as a POW, he painted attractive young women who bore a strong resemblance to my mother, including the heroine of his cartoon strip, "PENNY." Thus, the subjects for his oil painting had shifted!

He also practiced his hobby of photography, not only taking photographs with his collection of cameras but also developing the photographs himself. I remember my mother talking about how he would "take over the bathroom" by setting up his own darkroom! He probably also perfected his skills in building models from scratch. My mother said that he would abscond with her emery boards, which my father called "sandpaper sticks"—just perfect for sanding down those hard-to-reach areas.

My father's work at PAFB—being on the ground floor of a new missile program and working at a new Air Force base—was very rewarding. The living arrangements were great in my parents' beach house, their social life was vibrant, and Ben's Air Force career was off to a good start. My parents happily celebrated their eleventh anniversary on December 6, 1952, and what an interesting and eventful eleven years it had been!

However, in early 1953, another adventure was about to begin. My mother was pregnant!

Chapter Twelve
Nancy Jan

My parents had wanted a child for a long time, so they were ecstatic about the possibility of a new baby in the family! My mother wanted a baby girl with blue eyes and the name "Jan."

Ben continued with his development and testing of the Matador while my mom was busy building a nest for me and taking good care of her health. She purchased some adorable fabric with a clown and circus theme and made curtains for my nursery. Daddy Ben's skills at making models, photography, and painting portraits took another twist as he hand-painted my nursery furniture to match the fabric (Figure 12-1).

Figure 12-1: Jan's nursery with chest made by Daddy Ben and circus paintings by him to match curtains

The night before I was born, Mom went to a Halloween party and dressed as a farmer's daughter! On the morning of Sunday, November 1, 1953, my mother was reading the Sunday comics in the newspaper when her schedule changed. She alerted my father that I was coming, and he quickly drove her to the PAFB hospital.

I was a big baby—eight pounds and thirteen and a half ounces—especially for my small-framed mother. She had the blue-eyed daughter she wanted and named me "Nancy Jan." I came home with my parents a week later on November 8, and their lives would never be the same!

Figure 12-2: Daddy Ben and two-month-old Jan, Christmas 1953

In our beach home, the new family of three settled in. For my first Christmas, I was showered with gifts, and we had a big Christmas tree. One of my favorite pictures of me with my father is shown in Figure 12-2, when I was almost two months old. I was probably one of the most photographed babies on the planet, as my father took lots of pictures during my first year and developed them in his own darkroom. We could fill this book with baby pictures from that time, but I will only include two of them (Figures 12-3 and 12-4) so that you can see what I looked like with my triple chin and the rolls of baby fat on my legs!

I don't remember any stories about my first year, but from all indications, I was a good baby. Daddy Ben wrote in a letter to me on my first birthday, "For a year now, we have watched you grow and develop, and you have never ceased to amaze us, Jan. You have never complained, and you have asked for so little. Your little heart has already shown enough courage and understanding to be a credit to a person older than I. You have been so happy."

Figure 12-3: Seven-month-old Jan and her mother, Dolly Jo Smotherman

Figure 12-4: Eight-month-old Jan

My father continued to fly at PAFB, and years later, I would also fly in and out of PAFB in the T-38s that we astronauts flew. I even visited the hospital there, where I was born. I had come full circle—being born on the base where I would fly some thirty-five years later. Just down the road, north of PAFB, is the Cape Canaveral Space Force Station (formerly Air Force Station) in which the Kennedy Space Center (KSC) is located and where my father used to test the Matador. It was there that I would spend many hours training and would ultimately launch from the Space Shuttle launch pads on the cape jutting out into the Atlantic Ocean.

As the Matador and other missile programs grew, the Matador was deployed to support the North Atlantic Treaty Organization in West Germany where testing and development occurred. So, in 1954, my father was assigned to Germany to work on the development of the Matador systems. This seems ironic to me, as the German Air Force had been shooting him out of the sky and holding him as a prisoner about ten years earlier. Nevertheless, Germany is where the USAF assigned Daddy Ben to go, and he had orders to be there before my first birthday. Therefore, my mom and I spent my first birthday in Texas while my father was based at Ramstein Air Station in Germany. A picture of that first birthday is shown in Figure 12-5.

Figure 12-5: Jan's first birthday (in Fort Worth, Texas, with her mother); Daddy Ben had already deployed to Germany, November 1954

One benefit of my father being on temporary duty during my first birthday is that he wrote a very long letter to me. I have always treasured that letter, and I love to reread it. He wrote well, and it is a beautiful message.

Here is a snippet:

Tonight you are one year old. All too soon, you will learn how much courage it takes to be happy in this world. All too soon, you will find that you can't have all that you want. There will be disappointments, big and small, and there will be times when you will think your heart will break. But it won't, Jan, it won't. You have the courage to be happy. It really doesn't take much to be happy, honey, but it does take courage. More often than not, you will find that you already have what you really want. It isn't always the big things in life that count but the little things that are so often overlooked in the mad scramble of life. You will be ambitious, as I want you to be, but don't lose sight of what you have at all times. Keep the courage that your mother has given to you. Treasure the happiness that your mother will guide you toward.

This perspective on life is one that Daddy Ben must have experienced after being a POW for twenty-two months. He learned what is important in life and what happiness and contentment truly are. He thought about this a lot when only his basic needs were given to him, yet he still found happiness and contentment because he knew what was important. Furthermore, my mother had been courageous during my father's two years in combat and time as a POW. My father was trying to pass along to his one-year-old daughter these lessons that they had learned.

According to my passport, my mother and I traveled to Germany in December 1954 and were able to spend Christmas there with my father. Resplendent with a big Christmas tree and lots of presents, my second Christmas was a production! Like any precocious one year old, I took it all in.

Daddy Ben was first assigned to Ramstein Air Station, but after my mother and I arrived, he transitioned to Landstuhl Air Base in late 1954. We lived in the housing on this brand-new base, and I spent my toddler years there. A picture of me on the base with the officer housing in the background is shown in Figure 12-6. Some of my first memories as a child are small details of life in Germany.

It must have been hard for Daddy Ben to return to Germany, to hear German being spoken, and to see the results of the bombing of Germany that had occurred when he'd flown in WWII. He had unpleasant memories of his time in prison camp there, the food, and the attitude of some of the prison guards and officers. He never wanted to eat sauerkraut, brown bread, or blood sausage again. Luckily, as a USAF officer, he had plenty of good American food on the base and could get groceries in the commissary.

Figure 12-6: Jan (one and a half years old) in Germany, where her father was stationed

Figure 12-7: Jan's second birthday party in Germany, November 1955

Meanwhile, my mother was busy being a Gray Lady Red Cross volunteer in the hospital, playing bridge with her friends, and traveling all over

Europe. My parents had an active social life with lots of parties and events. I have pictures of my second birthday party with my friends, complete with a big cake, presents, and all of us sitting around a little card table (Figure 12-7). I also have pictures of my Halloween costumes and other American traditions as well as various activities in our home on the base.

My mother had a core group of friends she maintained for many years, even after we left Germany. She traveled with these women, and their children were about the same age as I was. I still have mementos from Mom's trips—most notably a silver charm from each place she visited in Europe. Probably my favorite of those charms is one she bought in Oberammergau, Germany—the site of the Passion Play (which I attended in 2022). It is a cross with the risen Christ's robe draped over it, and I wore this charm on a necklace when I was baptized. I carried this cross on the Space Shuttle with me, as it was very special to me.

Mom also bought several beautiful pieces of handblown Murano glass in Italy, hand-engraved and -painted artwork in England, china and crystal in Germany and Austria, and a gorgeous hand-carved and -painted nativity set in Germany. Plus, she bought the cutest hand-embroidered Swiss and German dresses for me. One of my favorites is shown in Figure 12-8.

Figure 12-8: Jan (age one and a half) in a Swiss embroidered dress purchased in Europe

Meanwhile, my father was introducing the Matador and other missile systems to our partners in West Germany. He was instrumental in establishing a USAF tactical missile operational capability in Germany. The Matador guided missile, armed with nuclear warheads, was designed to be launched against the enemy in Europe.

There were dozens of launching platforms around the forests and fields of West Germany, each about the size of a football field. My father wrote an article, "Missiles and Men," for the *TAF Review,* the Twelfth Air Force magazine, that chronicled the early days of the Matador, comparing its development to the growth of a child. He described the Matador program after its move to Germany in the following manner: "A new atmosphere prevailed. People were thinking in terms foreign to mere training and testing. The missile men were as frustrated as any freshman on his first day at college… They learned that many techniques which had produced excellent results in Florida produced nothing but bitter failure in the damp German forest. They learned that they could not get through this school unless they first survived."

At Landstuhl, my father continued to pilot airplanes, including the C-47 and the T-33. Landstuhl was in the southwest part of Germany. In WWII, Daddy Ben had been flying bombers over the northern part of Germany. So, when he was flying in 1955 and 1956, it wasn't over territory that was familiar to him during the war in 1943. Nonetheless, it was assuredly an odd sensation to fly over Germany as friendly territory after bombing industrial targets in enemy territory. By the end of 1955, my father had accumulated a little over 2,082 hours as a pilot and was promoted to senior pilot on June 5, 1955.

Our family continued to live in Germany as the new year approached, and my parents celebrated their fourteenth wedding anniversary in December 1955. Their Christmas picture with little Jan is shown in Figure 12-9. While growing up in Germany, I was a stubborn and active girl with curly blond locks, and I loved riding my tricycle in the house as well as outdoors.

My parents missed their families in Texas for many milestones and holidays, and it was an adjustment to live in a different culture and country. Add to that the trials and tribulations of raising a toddler, and it was a difficult time. Still, while Mom traveled and socialized, Daddy Ben enjoyed his important job. He continued to fly C-47s, and through his assignment in Landstuhl, he finished up with over 2,267 hours of pilot time at the end of 1956.

As the development and qualification of the Matador missile continued, testing and training was conducted at Bitburg AFB and Hahn AFB in Germany. However, due to the large expanses of land needed to adequately test the Matador, each Matador missile was required to be qualified at Wheelus AFB in the deserts near Tripoli, Libya, from 1954 to 1956. Therefore, my father traveled to the Libyan coastline with its sparkling Mediterranean waters to test the Matador. Coincidentally, this base was used by the Luftwaffe in WWII during the North African campaign and was known as Mellaha at that time.

Figure 12-9: Smotherman 1955 Christmas card

These tests seem very exciting to me, and they were for my father as well. What a sense of accomplishment he must have felt with each test, and what a great experience it must have been to travel to this large American base in Libya and celebrate after a successful launch! The development of this missile technology in which my father was involved directly applied to the development of the rockets that were used by our country's space program. As my father witnessed these many Matador launches, and later other missile launches, I know that in the future he looked forward to seeing a Space Shuttle launch with his daughter on board.

A picture of my father with some models of the Matador is shown in Figure 12-10. In the conclusion of the article about the Matador, my father wrote: "It was known all along that these children of history would someday grow into manhood. So many times they felt like orphans, but they were never vindictive. So many times they were subjected to criticism, but they were not dismayed. Their schooling was fraught with failures, but they were not discouraged. They analyzed, studied, planned, and organized their thinking toward improvement and progress. Long before, they had learned that they must first learn to walk before they could run. There was only one way to do a thing, and that was the right way…

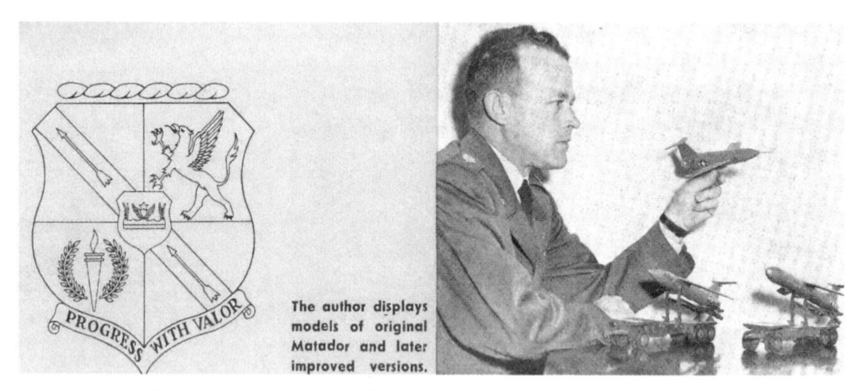

Figure 12-10: Major Ben Smotherman with models of Matador missiles from the "Missiles and Men" article he wrote for the Twelfth Air Force magazine, 1957 (USAF photo)

"The way of life that an individual develops is influenced to a very great extent by the impressions formed in early childhood. Integrity is developed early and is only strengthened in subsequent years. Experience, the great teacher, exerts pressure here and there in the process of molding character. If the individual learns a lesson from a mistake he makes, then the mistake has served a purpose. It is all the process of growth. That is why a parent derives such joy out of watching a child grow up."

I enjoyed reading this article and wanted to include it because it not only described the development of the Matador, but it also demonstrated the quality of my father's writing. I am proud of him for his technical ability and leadership in the development of the Matador as well as for his incredible gift of being able to communicate so well in his writing.

Following the travel and excitement of going to Libya for testing, there was trouble at home. I don't know what happened, but in 1956, my parents' marriage was falling apart, so my mother and I flew back to Texas in December 1956. What a lonely and miserable Christmas each of us had that year. The Smotherman and Gantz families in Texas wondered what was going on and tried to comfort my mother, my father, and me during this separation. I was only three years old, but I was sad and confused and missed my Daddy Ben.

In 1957, their divorce was initiated and processed. My mom went back to school to work on a graduate degree in psychology at Texas Christian University (TCU) and worked there as a graduate student. During the day, I stayed at Mrs. West's house in Fort Worth along with several other children, as she had daycare in her home. I have fond remembrances of Mrs. West and my fun times with the other children under her care. We played outside a lot, and I remember that whenever a plane flew overhead, I would point at it and proclaim, "There's my daddy!" After all, I was only three years old, and every plane that I saw flying was my daddy's! Maybe that is where I learned to always look up at the sky whenever a plane flies overhead, but it is a fond memory.

In 1957, my father stayed at the Landstuhl base and continued to work on the Matador program. It was tough and lonely for my father in Germany. Likewise, it was difficult for my mother to return to the United States with a toddler in tow and an uncertain future. By the end of 1957, after sixteen years of marriage, the union was officially over.

Chapter Thirteen
Two Families

By the end of October 1957, the Matador was operational in Europe and ready for whatever war it was called upon to serve. The knowledge from this program was used to develop and test other missiles and rockets for the United States, so it was time for my father to use his experience for other programs.

In November 1957, as I was turning four, Daddy Ben transferred to the Cooke AFB north of Los Angeles, California. On his way there, he no doubt spent some time visiting me as well as his parents and other family members in Fort Worth.

In 1957, my mother and I spent our first Christmas together as a family of two. I was a curious child who enjoyed taking things apart and playing with toys such as Lincoln Logs, Tinker Toys, etc., that allowed me to build things (Figure 13-1). I guess the budding engineer was evident when I took apart a doorknob when I was three or four years of age! I heard this story from my mother throughout my life, but she never told me whether I put the doorknob back together again.

With the advent of the missile age in the 1950s, of which Daddy Ben played a significant role, there arose an urgent need for a combat-ready missile base, complete with training capability, for the Air Force. My father was one of the first Air Force officers at Cooke when this organization was activated. Once again, Daddy Ben was at the forefront of this effort to transform Cooke AFB and develop missile operations and training. He had the perfect background and experience to take on this new role, and it was exciting for him to embrace this challenge in this relatively new field of technology.

Figure 13-1: Jan enjoyed toys where she could build things and take them apart. Christmas 1958, five years old

Also, about this time, the Russians were gaining ground in the field of launch vehicles, which had been adopted from their strategic missile technology. The launch of the Russian Sputnik 1 satellite into orbit on October 4, 1957, followed a month later by Sputnik 2 that carried a dog, Laika, into orbit, had military implications and caused an immediate acceleration of the USAF missile program.

Wow, talk about timing! Daddy Ben arrived in California for his new assignment as the U.S. missile force was growing by leaps and bounds due to the race with the Russians, and the U.S. was launching missiles from Cooke. My father was a part of the space race! At Vandenburg, they eventually managed, trained for, and operated launches of the Thor (whose responsibility had moved to the Air Force from the Army's Redstone Arsenal in Huntsville, Alabama), the Atlas, and the Titan missiles/rockets.

It is interesting to note that Vandenburg AFB was also the site selected for Space Shuttle launches for the Department of Defense. Although the infrastructure was in place for these launches, the demise of the Space Shuttle Challenger in 1986 canceled the plans to launch from Vandenburg. While working for NASA as an executive, I visited Vanderburg AFB to witness the launch of the Gravity Probe B scientific satellite, and I also went there occasionally for meetings. Today, Vandenburg is part of the U.S. Space Force and is called Vandenburg Space Force Base. Many uncrewed space missions are launched from there.

In California, Ben was off to a new start in his career and in his personal life, but he was alone. He always loved dogs and horses, so it is not surprising that, as a brand-new bachelor, he pursued his lifelong dream of owning a horse. He bought a beautiful palomino and, along with it, all the necessary tack—the western clothes, chaps, and boots essential to look like a real cowboy (Figure 13-2)!

Figure 13-2: Ben Smotherman and his horse in California, 1958

It is fortunate that my father had lots of hobbies, and he now had some time for painting and drawing. The subjects of his paintings became western in nature, and he painted many scenes that included horses and cowboys (Figure 13-3). He also designed stagecoaches and built them from scratch. These took years to complete, but he used the time in California to perfect his techniques. I have several drawings showing the details of his designs, the history of each coach, and how each model coach would be built.

In 1958, in addition to his demanding job, hobbies, and time riding his horse, Daddy Ben also continued to fly airplanes. He flew C-47s exclusively and was able to squeeze in a few hours of flying every month. He also received great news in April 1958 right after his thirty-ninth birthday: He had been promoted to lieutenant colonel! I hope that he had lots of friends and colleagues who could help him celebrate the occasion.

Figure 13-3: Ben Smotherman further developed his painting skills with Western oil paintings while in California (1958 – 1959)

At the same time, I was still going to Mrs. West's every day and learning things that a four-and-a-half-year-old child learns about books, play, friends, and life. I remember seeing my mother's parents, my Mono and Dado (Gantz) in Dallas, a lot more than I had seen them earlier in my life. I also was able to see my aunts, uncles, and cousins who lived nearby. I don't remember being sad; I had a full and happy life.

During 1958, my mom was continuing graduate school and working. At the same time, she started dating Bryce Davis, who was also a psychology student working on his graduate degree at TCU. Later in the year they married, and I gained a stepfather whom I called "Dad" (Figure 13-4). Whenever I mention "Dad" from now on, that is to whom I am referring.

In 1958, another important milestone in my life occurred—NASA was created by President Eisenhower on July 29. NASA was set up as a government agency that would focus our nation's efforts on a civil space program, aeronautics research, and space research. Still reeling from the Russian launch of the Sputnik satellite in 1957, the U.S. formed NASA to conduct all non-military space activities.

Figure 13-4: The Davis family—Dolly Davis, Jan, and stepfather Bryce Davis (1959)

In Fort Worth, Texas, I progressed from daycare at Mrs. West's to kindergarten. I have always loved learning and school—apparently even at that young age. When I had just turned five years old, one of my mother's letters to my father in 1959 said, "At Christmastime, the school had open house, and the children sang songs, presented their mothers Christmas cards they had made, exchanged gifts, etc. Her schoolwork has been very rewarding, and she has learned so much… Her teacher is a refined, elderly woman and is well-trained." That teacher was Mrs. Aslakson, with whom I continued visits and correspondence until she passed away.

In August 1959, my father transferred to the Pentagon in Washington, D.C. He was in the USAF Directorate of Plans and Operations where he was active in strategic ballistic missile planning during the force build-up. Of grave concern in the late 1950s and the early 1960s was the development of the intercontinental ballistic missiles (ICBMs) by the Soviet Union, and my father was involved with the strategies that our country would use to deploy and operate our own missiles if an attack was warranted. He was also heavily involved in Joint Chiefs of Staff deliberations and congressional hearings on U.S. and foreign ballistic missiles.

Meanwhile, my mother, my stepfather, and I moved from Fort Worth to Waco, Texas, after Dad completed his master's degree in psychology at TCU. In the fall of 1959, I enrolled in first grade at a private school there called the Duncan School for Childhood. I loved going to school, and I adored my teachers as well as the beautiful white Victorian building that housed the school. My favorite subjects at the tender age of six were reading and music.

The start of a new decade in 1960 brought hope for happiness for my father, my mother, my stepfather, and me. The 1950s had been full of growth and opportunity but had also offered disappointment and loss. For example, when my father moved to the booming metropolis of Washington, D.C., he lived in Alexandria and was consumed by the responsibilities and pressures of working at the Pentagon. It was very different from the wide-open desert in California, and he was not able to bring his horse with him.

Fortunately, he was able to make time for flying, socializing, and his hobbies. Plus, in 1960, he met a woman named Jacqueline (Jackie) Bostic, and they were married. The next year, they welcomed their firstborn, Darby Ann, who was named after Ben's sister's married surname, Darby. Once again, Ben had a family and a baby around the house.

My father's life was full of transitions in the early 1960s, as his flying days ended on September 23, 1961, and the last airplanes that he flew were the C117 and the U-3A. He ended the flying portion of his career after flying in many different aircraft, mostly C-47s and variants of the B-17. From my flying days at NASA, I learned that the last flight of an airman, known as a final or "finny" flight, was celebrated with a fountain of fire extinguishers or bottles of champagne that totally doused the pilot! I hope that my father's finny flight was dutifully celebrated and that he was soaked while among friends and fellow fliers, even though he was undoubtedly a little melancholy.

Daddy Ben finished his flying with a total of 2,521.4 hours as pilot and 225 hours as a student for a total of 2,746.4 hours. Of those, he flew a total of fifty-five combat hours. His boyhood dream had been to fly airplanes, and he'd worked his way through school, worked to pay for flying lessons, and continued to work hard while learning to be a pilot in the Army Air Forces. He'd flown for over twenty years, and I am thankful that he enjoyed flying so much that he wanted to continue, even after his combat duty. After all, many of the pilots who flew in combat did not ever want to fly again, either commercially or as part of the military.

Transitions were also in order for me and my family. After my completion of first grade at Duncan, my mom and stepfather had an opportunity to move to Florida and work at the Northeast Florida State Mental Hospital. My mother always wanted to live near the ocean, so they moved to a small community, Macclenny, just west of Jacksonville. We lived in a concrete house with jalousie windows on a corner near downtown Macclenny. It was perfect for enjoying the Florida weather while relaxing on the porch.

During my formative years in Macclenny, Florida, I created lots of memories. I continued to read voraciously and developed a fondness for reading biographies because I wanted to understand how some people developed their fame or discovery. I especially remember reading about Marie Curie and other scientists and being fascinated by all things science. Biographies are still my favorite genre, and that is one motivation for writing my own biography—that it might inspire someone to pursue their dreams or discover new things.

Due to my excitement about reading books, I had a large personal collection of children's books. My mother and a friend, Emily Taber, worked hard to establish a library in Macclenny for Baker County. I donated my books to the library, and those books became the children's section. The Emily Taber Public Library still exists today. It's much larger now and has its own charming building in the historic courthouse.

I spent second, third, and fourth grades at Macclenny Elementary School. In addition to reading and spending time outdoors, I also developed keen interests in crafts, sewing, animals, and riding my bicycle. In such a small town, I could ride my bike anywhere, and it was how I got to school every day. It was a simple and joyful life, and I remember riding my bike downtown to spend my allowance at the five-and-dime store. There was one red light in town, and the community was very safe.

As soon as we moved to Macclenny, I wanted a pet dog, and when I was about nine years old, we adopted the cutest little beagle. I named her "Flopsey" for her floppy ears. She is shown in Figure 13-5 alongside me and my trusty bicycle. Flopsey and I went everywhere together, and my affection for animals was cemented forever.

Figure 13-5: Jan and her dog Flopsey in Macclenny, Florida, where she spent second, third, and fourth grades

My love of the outdoors and an active life meant that I also swam every chance I could get at the public pool across the street from our house and played outside until the sun set or I was called home. My friends and I climbed trees; told stories; and played tetherball, badminton, and whatever we could find. What a great life we had!

My mom and I wanted to form a girls' club since there were no Girl Scout or Camp Fire Girls troops. So, we started a group with seven of my friends and called ourselves the Jolly Eights, as there were eight of us and we were eight years old. I was elected the president, and our club colors were red, white, and blue. I wanted the club color to be lavender (a mixture of red, white, and blue), but I was overruled, and I remember pouting about that! That was one of the first lessons I learned about leadership and teamwork.

We went on field trips, did crafts, played games, and had lots of fun. I enjoyed the creative process, whatever it was that I was making, and I especially enjoyed projects that involved textiles or sewing. My mom also liked the needle arts, and she taught me how to sew, embroider, knit, crochet, and needlepoint. Although she taught these crafts to me and was obviously very talented, I often wonder if I also inherited my affinity for crafts from my father. Even though I was not around him very much, I believe he must have been an influence on my artistic ability as well as my penchant for engineering.

For our three years in Macclenny, I had a happy, full, and multidimensional life. Almost every weekend, we went somewhere in Florida or went to Jacksonville Beach for the day.

Meanwhile, my father and his new family were growing and getting established while he worked at the Pentagon. He and I continued to communicate, but the distance between us was not conducive to seeing each other.

Little did I know that my mom and dad had been trying to have another child. They were successful, and our lives were changed forever when my brother Ronald Addison was born in 1962. At nine years old, I was the best babysitter and caregiver that you could imagine (even though I wanted a sister!) I did not meet my other siblings for about ten years, so I did not have that same great experience watching them grow up as I did with Ron.

On May 5, 1961, Alan Shepard was the first American in space, and twenty days later, President John Kennedy made his famous speech about landing a man on the moon by the end of the decade. The wheels were set in motion for the boom in Huntsville, Alabama, where the engineering and testing of the Saturn V rocket was conducted.

One of my stepdad's colleagues from the mental hospital, Dr. Himon Miller, had moved to Huntsville, and he asked Dad to join his private practice. The timing seemed perfect to move an inquisitive nine-and-a-half-year-old girl and a newborn baby to a new home with new opportunities. So, in the summer of 1963, the four of us made the transition to Huntsville, Alabama.

Dad joined Dr. Miller's practice and was pleasantly surprised to be the first psychologist in Huntsville. As Huntsville was growing so fast, he decided to open his own private practice, with my mom (who, remember, had a degree in industrial psychology) serving as his bookkeeper and psychometrist, administering standardized psychological tests to his patients.

My father and his new family had also grown by one more, as my sister Holly Elizabeth had been born in 1962. What joy and activity changed his life, as he now had one-year-old and newborn daughters! It was a new beginning for two distant but connected families.

Chapter Fourteen
Rocket City

As my stepfather was pursuing a new career in Huntsville, Daddy Ben was being assigned a new job as well. In the summer of 1964, he moved to Manchester, New Hampshire, to become the commander of the Air Force's New Hampshire Satellite Tracking Station and Grenier AFB.

While adjusting to a family again and having a pressure-packed job, he was delighted with the birth of his son, Benjamin (Benjie) Charles, in 1965. Daddy Ben's life changed a lot in those few years, and he enjoyed being a father to three children at home. A picture of him giving a doll to Darby and Holly is shown in Figure 14-1. Unfortunately, I was not able to be around these new half-siblings, as time and distance separated us, so I did not get to know them until much later in life.

Figure 14-1: Daddy Ben showing a doll to his daughters Holly and Darby (Christmas 1964)

My first impressions of Huntsville, Alabama, were positive. What had started out as a sleepy cotton town was nestled into the rolling foothills of the Appalachian Mountains (Figure 14-2). Along with its newfound mission of a nascent space program, it comprised beautiful antebellum homes that had been preserved. The Tennessee Valley Authority had dammed up several portions of the Tennessee River into lakes, thus creating a recreation mecca of fishing, camping, water sports, and sheer beauty. It was a wonderful place to live, and as my family were outdoors people and very active, we enjoyed the area.

Figure 14-2: The foothills of North Alabama, where Jan and her family moved in 1963

Shortly after President Kennedy's announcement on May 25, 1961, that we were going to the moon, the NASA centers hurriedly prepared facilities, technology, and personnel for this momentous goal. It was during this time of extensive growth in Huntsville that my family settled there in 1963. By that time, NASA was finishing up the Mercury missions, which had one astronaut on each mission, and beginning the Gemini program with two astronauts on each flight. It was clear to me that Huntsville was a very special place and played a key part in our nation's space program. The main reason was that the Army's ballistic missile program evolved into NASA rockets that took astronauts into space. Leading that effort was Dr. Wernher von Braun and his team of scientists and engineers.

Figure 14-3: Jan and her brother Ron on his first birthday, shortly after moving to Huntsville, Alabama, where they lived in an apartment for a year due to rapid growth in Huntsville from the booming space program (August 1963)

Growing up in Huntsville in the 1960s was a fascinating and educational experience. I was enrolled at the private Randolph School in fifth grade, right after my brother's first birthday (Figure 14-3). As Randolph was a small school, I had small classes and made fast friends there. Many of my friends were children from Dr. von Braun's team, including his daughter, Margrit. Most of my friends' parents worked on the space program, and they fueled my interests and kept me engaged with our journey to the moon as we began the Apollo program. This was a new world to me, and at an early age, I learned to appreciate other countries and people with different backgrounds.

As the space program progressed, the Saturn V rocket that would carry three men to the moon was being developed at the Marshall Space Flight Center (MSFC). Each engine in the Saturn V was tested in Huntsville, and we could hear (and feel) the roar of the rockets and the vibration in the ground. In fact, when the five-engine cluster for the first stage of the Saturn V rocket was tested, it was incredibly noisy, and sometimes the vibration would break windows around town. We knew that a successful test meant one more step to the moon, and Huntsville would be a part of it. It was exciting for the whole community, and while going to school in Huntsville in the 1960s, whenever there was a launch, the teachers would wheel a little black-and-white television into the classroom so we could all watch it on TV.

My other activities in Huntsville in the fifth grade initially included swimming, church, and Girl Scouts. I enjoyed my Girl Scout troop, where I started as a Junior, and I am a big fan of Girl Scouting to this day. I continued to develop my skills in sewing and needlework, and along with Mom and her trusty Featherweight sewing machine, I took great delight in making clothes and other things.

In the summers of 1964 and 1965, I went to the local Girl Scout camp on Lake Guntersville, called Camp Trico (for "Three Counties"). It was a game changer for cultivating my love of the outdoors. Camp Trico was on a beautiful part of the lake, and I learned to canoe and worked even more on my swimming skills. I also took up archery, and when I came home, I wanted to buy a canoe! My parents decided that a motorboat would be more practical for the family, so they bought a fifteen-foot tri-hull boat with an outboard motor. Dad could go fishing, and I learned about boating, skiing, and enjoying the lake life. Boating on Lake Guntersville was a popular activity for Huntsvillians, and Dr. von Braun even had a boat—a Chris Craft named Orion. As shown in Figure 14-4, he often took his family and friends to the lake for a respite from the hectic race to the moon.

At about the same time, my father was well into his tenure as the commander of the New Hampshire Satellite Tracking Station and Grenier AFB. On my eleventh birthday, his letter to me was written on the letterhead, "6594th Instrumentation Squadron, 6594th Aerospace Test Wing, Air Force Systems Command, United States Air Force Grenier Field, Manchester, New Hampshire." From 1964 to 1967, as a colonel there, he managed over six hundred engineers and technicians who were engaged in continuous, twenty-four-hour daily operations tracking Air Force satellites. A picture of him while commander of Grenier AFB and the New Hampshire Satellite Tracking Station is shown in Figure 14-5. I'm so glad that Darby and I were able to visit the station recently (now New Boston Space Force Station) and retrace his footsteps.

Figure 14-4: Wernher von Braun enjoying his Chris Craft Orion on Guntersville Lake in 1955. Pictured are daughter Iris; wife, Maria; daughter Margrit; Dr. von Braun; friends Irmgard and Ernst Stuhlinger (holding son, Tillman, with daughter, Suzanne, barely visible); and colleague Heinz Koelle (photo courtesy of U.S. Space and Rocket Center Archives)

Figure 14-5: Ben Smotherman as commander of the Grenier AFB and the New Hampshire Satellite Tracking Station (USAF photo)

When Daddy Ben was there, the station had one satellite-tracking dish and was the first tracking station to attain the capability to track two satellites simultaneously. There was a lot of history at the base. It had many munitions and bombs in the lake, as it had been used for bomber training in WWII. It also had ruins of a residence that had been used in the Underground Railroad. Daddy Ben enjoyed playing a critical role in the community and in the conservation of the environment in his surroundings.

While in New Hampshire, Daddy Ben went to college at the nearby New Hampshire College (now the University of New Hampshire at Manchester). He had always wanted to earn his college degree, and in June of 1967, he graduated with a bachelor's degree in business administration.

In New Hampshire, my father enjoyed the outdoors in the beautiful countryside of his roundabout. He also continued his hobbies of woodworking and carriage making when he had the time. His office was decorated with very intricate (and working) models of coaches that he designed and made from scratch, such as the Wells Fargo wagon, the 1902 Standard Oil Gasoline Delivery wagon, the 1870 Concord Stagecoach, and the 1864 Lumber and Civil War Cannon wagon. A drawing of one of his designs is in Figure 14-6, and one of these exquisite coaches built to one-eighth scale is shown in Figure 14-7. During this time, Ben also continued his hobbies of photography and oil painting and worked on his golf game.

Seventh grade in the fall of 1965 was a formative period for me. The fifth through eighth grades is when girls generally decide what their favorite subjects are. For me, I was fortunate to have an excellent science teacher in Mr. Herbert Walker. He was one of those teachers who pursued education because he loved to teach and to introduce youngsters to the wonders of science. He taught college-level chemistry and physics to us as well as instructed on how to write in a laboratory notebook, how to work in a scientific laboratory, and much more. In addition, the math program at Randolph was an accelerated one, and I enjoyed my math and algebra classes almost as much as my science classes.

For my science project on gyroscopes, my parents called upon one of their friends, Dr. Joyce Neighbors, who was an engineer at NASA. Joyce was able to borrow a portable electronic gyroscope from von Braun. Of course, I demonstrated the basic science and physics of a gyroscope with a simple toy version, but the electronic gyroscope was much more powerful and was truly amazing to me. I also thought that it was special to have the gyroscope that von Braun used! It was a brilliant strategy for my parents to have a woman engineer mentor me. She was the first woman engineer whom I had ever met, and she was a role model for the rest of my career.

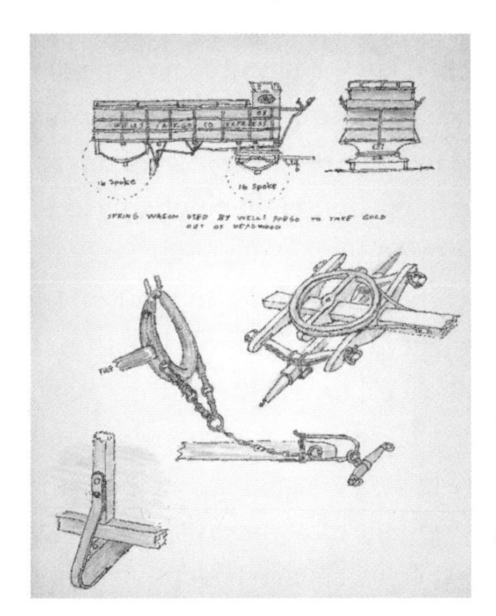

Figure 14-6: Drawing of one of Ben Smotherman's coaches; he built every part from scratch after much research

Figure 14-7: One of Ben Smotherman's handmade coaches (1/8 scale) built from scratch and complete with moving and functioning parts

Taking all this into account, in junior high, I had a bearing on where my career would go. I had teachers who'd greatly influenced me at that time, particularly Mr. Walker. I also had a relatable role model who helped me understand what engineering was and how it could be applied. Plus, I had a newfound interest in space! It was no longer just about rocket engines that were tested in Huntsville and shook the ground or even about rockets shooting up into the sky. The fundamental laws of physics, the application of engineering to cool projects, and the scientific discoveries made by the space program were being revealed to me. The progress toward our country's goal of landing on the moon was happening right before my eyes in Huntsville, and it was considerably inspirational to me.

Shortly before my graduation from Randolph, in the fall of 1967, my father was sent to his last assignment in the Air Force, which was at the Foreign Technology Division of the Air Force Systems Command at Wright Patterson AFB in Dayton, Ohio. Here, he was responsible for the management of all analyses and assessments of aerospace scientific and technical intelligence programs covering all foreign aerospace systems and related technology. As part of his job, he managed over five hundred professional personnel, largely scientists and engineers, and was responsible for inputs to the national intelligence programs.

I did not see Daddy Ben much when he was in New Hampshire or Ohio, but we frequently wrote long letters to each other. My mom also sent pictures to him from the newspaper or school paper about my activities. He was always very encouraging and proud of me and wrote the sweetest messages.

My other activities growing up in Huntsville were piano lessons and ballet classes mixed with a little softball and cheerleading. Plus, one of my best friends, Gigi Luehrsen (the daughter of one of von Braun's team members), invited me to go ice skating with her on a Friday night in 1965. I loved it so much that I went again on Saturday and Sunday. Huntsville did not have a regulation-sized rink, but Ben Wilcoxon, the father of a friend of mine, was an engineer/entrepreneur and wanted to have an ice rink, so he built the first and only one in Alabama for fourteen years.

My instructor/coach was Robert Unger, a German who lived in Knoxville and traveled to Huntsville every other weekend. He was a strict disciplinarian—no dirty skates, no chewing gum, and punctuality for lessons were the rules. I learned so much from Mr. Unger besides ice skating. He taught me the meaning of discipline, hard work, and trying repeatedly until I got it right. A picture of me skating in one of Mr. Unger's ice shows is in Figure 14-8.

Figure 14-8: Jan Davis ice skating in an ice show at the Ice Palace in Huntsville, Alabama (1970)

I took many skating lessons—free style, dance, synchronized skating, and figures (like figure eights). My mother grew tired of driving me to these lessons and waiting for me all the time, so she decided to take ice skating lessons too. With her athletic and dance background, she became quite good in a short period of time. In fact, she soon became a skating instructor for the little ones. In this role, she received free admission to public skating sessions, so she and I went skating every night (once I finished my homework). We became very close, and ice skating was something that we both loved and had in common.

In 1968, my studies wrapped up at Randolph, and I graduated from ninth grade. I received a rich education and was thankful for learning French as a foreign language as well as developing my affinity for math and science. My graduating class only had twenty-seven people, and I was as close to them as I was to my other friends and teachers. Randolph did not have a high school then, so it was time to move on to Huntsville High School.

In the late sixties, it was an exciting and impressionable time in my life as the three-person Apollo missions marched on. In 1969, my parents decided the Apollo 11 launch in the summer would be a historic event, so we should go to Florida and witness it. They had some friends who were temporarily living in Florida while working at KSC, and we stayed with them during our trip.

On the way there, I made two signs for our Ford Galaxy XL. One said, "Look Out Moon, Here Comes Huntsville!" and the other said "To the Moon or Bust!" In viewing the Apollo program, most of the world thought of KSC and Johnson Space Center (JSC) as the NASA centers that played the largest roles. Even at fifteen years of age, I realized that Huntsville did not get enough credit for its contribution to the Apollo program, so I made the signs to get everyone's attention!

After having a Shirley Temple and dinner at Bernard's Surf in Cocoa Beach, we went to bed early to be fully rested for the big day. Early on the morning of July 16, 1969, we were one of thousands of cars lined up on the causeway to watch the momentous event. It was estimated that one million spectators viewed the launch of Apollo 11 from the highways and beaches in the vicinity. There was excitement in the air as we all proudly waited for the countdown.

The launch was spectacular, with the big Saturn V and its cargo and crew slowly lifting off the launch pad—much more slowly than I had envisioned. A few seconds later, we could feel the rumble and hear the noise coming from the five F-1 engines that had been tested as a cluster in Huntsville. In response, the whole crowd erupted in shouting and applause. It was a unifying and wonderful feeling to be a part of that historic moment. Although I was already interested in NASA and our space program, watching that launch kindled a new passion for spaceflight that I have never lost.

We drove back to Huntsville with a newfound interest in Apollo 11 and followed the mission closely. On July 20, I was one of millions of people around the world who watched Neil Armstrong's first steps on the moon. It was an unbelievable achievement that our nation was able to accomplish, and one that I appreciated more and more in my adult life as I became deeply involved in the space program. I was relieved and elated that everything had worked out with safely landing Americans on the moon and returning them to Earth. In addition, the pictures of the Earth from the moon were incredible and forever changed our perception of the "blue marble" that we call home. I looked forward to Americans landing on the moon again.

As an astronaut, I had the opportunity to meet and interact with many members of the astronaut corps who had flown Mercury, Gemini, Apollo, and Skylab missions. Two of the people whom I came to know were Buzz Aldrin and Mike Collins. It was a thrill to share with them what an inspiration they were and how meaningful their Apollo 11 launch was to me.

After the mission, there was a big celebration in downtown Huntsville, and Dad's jazz band, "The Southern Comforts," played there. The crowds were enormous, and von Braun was hoisted on the shoulders of members of the crowd. We were all proud to be a part of the successful Apollo 11 mission, and we were proud of the German team from Peenemünde who now called Huntsville home. Von Braun made a famous speech with a memorable quote: "My friends, they were dancing here in the streets of Huntsville when our first satellite orbited the Earth. They were dancing again when the first Americans landed on the moon. I'd like to ask you don't hang up your dancing slippers."

After the Apollo 11 mission, I began my junior year in high school. Along with that, I began the college application process and started visiting colleges. Largely due to the influence of the space program and the engineers in Huntsville, I decided to pursue engineering in college. I was discouraged by many who said, "engineering is a man's field," but I was too stubborn to listen to them! My parents were very encouraging, and my father wrote to me:

I am particularly pleased that you have chosen to be an engineer, and I know that you will be a good one. I sense the same creative drive in you that always possessed me. It will take you to the top. Not too many years ago, a woman engineer was severely limited in what she could do, but that is not the case today. Any field of specialization is open to you, and you can compete successfully with your male contemporaries. With your beauty and charm, I'd say that you have a decided edge!

The summer of 1970 between my junior and senior year was again full of swimming, playing tennis, and ice skating. Plus, I worked as a lifeguard at a neighborhood pool. In the evenings, I took calculus classes at the University of Alabama in Huntsville (UAH) and made good grades. I was the first high school student to be able to take courses at UAH, and this is now a well-established program called the dual enrollment program. I am glad to have paved the way for so many other deserving students.

During my senior year, I was delighted to be accepted into the University of Tennessee and Georgia Tech. I realized I wanted to go to Tennessee so I could ice skate with Mr. Unger, and not because of its engineering program. I decided that I wanted to go to a smaller university that specialized in technical fields. Furthermore, being able to ice skate with Mr. Unger was not a very sound reason for attending a college. Therefore, I accepted the offer to attend Georgia Tech.

Overall, my senior year of high school was uneventful yet very busy. I concentrated on my studies and completed three more quarters of college calculus with good grades. My other activities included swimming, ice skating, sewing, and spending time with my family at the lake.

As Daddy Ben was winding down in his career, I was just beginning my path forward by graduating from Huntsville High School in 1971. It was a time of transition for both of us, with new opportunities and hope for the future.

Chapter Fifteen
An End and a Beginning

The summer of 1971 was fun, as I was preparing to go to college, working as a lifeguard, and teaching swimming. By the fall, I was ready for my collegiate career! I packed up my parents' car with clothes, bedding, and other things needed for my dorm life at Georgia Tech and started my new journey. I didn't know any women who were students there, and when I was a freshman, there were only about two hundred women out of a student body of eight thousand! It was a fresh start and a challenge, and I was willing to charge ahead into my freshman year.

At the same time, my father, a colonel in the USAF, was preparing for his retirement. In 1971, after thirty years of service, he was required to retire from the Air Force. As he and my mother had loved living in Florida when I was born at PAFB, he bought a house in nearby Satellite Beach, Florida. The house had a swimming pool and backed up to a canal, where he kept a boat. It was the perfect place to retire with three small children. In his new leisure time, he focused on hobbies such as golf, photography, oil and watercolor painting, wood carving, model building, drawing, writing, sailing, and flying!

The letters between my father and me increased when I went to college. I was on my own now, so I took advantage of that situation and had more communication with him. He had more time in his retirement as well. In January 1972, he wrote:

I have missed you through all these years, Jan, and I never ceased to love you and pray for your welfare. Circumstances kept us apart, and I was unable to do all the things for you that I would have liked. I feel that there were times in your life when you were lonely for me, and I pray the days ahead will make up for some of it. I want to keep in touch with you now; I want to write regularly to you and visit with you time to time.

Daddy Ben moved back to Texas in 1973 after living in Florida for about a year. I later found out that his marriage was failing, and he took the responsibility to help raise his three children, but he needed the support of his family in Texas. His divorce was finalized in early 1974.

While I was at Georgia Tech, my focus was on school, but I was still very interested in the space program. In the spring of 1972, the crew of Young, Duke, and Mattingly completed the Apollo 16 mission. This mission was very special to me at the time because John Young was a Georgia Tech graduate, and his launch and mission were celebrated while I was there! It was exciting to see the Georgia Tech pennant that had been installed on the moon buggy and brought back to campus for all of the students, faculty, and alumni to view.

While in school, I majored in engineering, science, and mechanics. At the beginning of my sophomore year, Georgia Tech offered two special degrees in bioengineering; one was in electrical engineering, and one was in engineering mechanics. The idea of combining biology with engineering really appealed to me. Both degrees were in the School of Biology, and I could also pursue a minor in either electrical engineering or engineering, science, and mechanics. Therefore, I switched my major to biology with a minor in engineering, science, and mechanics. This meant that my course of study would be in biology, and my elective courses would be in engineering. It was a tough curriculum but one that I enjoyed.

Sometime in the early 1970s, my mother and I drove to Dallas, Texas, to visit her parents, my Mono and Dado. I asked Mom if I could visit Daddy Ben while we were there, and she agreed. We went to church in Fort Worth so that I could hear my uncle (Ben's brother Don) preach, and Daddy Ben was serving communion there. That was a joyous day! He asked if I wanted to see my brother and two sisters, and I was delighted to have the chance to meet them! A picture of Daddy Ben, my brother Benjie, and my sisters Holly and Darby is shown in Figure 15-1.

While I was home in Huntsville for the summer of 1974, I met Paul Dozier at church. He was a chemistry student at Auburn University. Paul and I dated that summer and continued to date during my senior year at Georgia Tech.

Figure 15-1: Daddy Ben with his three other children, Benjie, Holly, and Darby (Fort Worth, Texas, early 1970s)

Georgia Tech was not easy; in fact, it was demanding. However, my perseverance paid off when I earned my Bachelor of Science degree in applied biology with a specialty in biomechanics in June 1975.

Paul and I were married in September 1975 in Huntsville. As he was still studying at Auburn, I moved there and decided to continue my education in engineering. So, I enrolled at Auburn University in mechanical engineering at the undergraduate level. As I had already completed so many engineering courses at Georgia Tech, it took a little over a year to earn my mechanical engineering bachelor's degree. It was the highlight of my college years, as I thrived in the small college town atmosphere.

I graduated from Auburn University in December 1977 with a Bachelor of Science degree in mechanical engineering. At that time, the best jobs were in the booming oil industry, and I wanted to work in research and development. My heart was still with the space program, but sadly, there was no human space program after Apollo, so there were no jobs.

I accepted a position with Texaco Exploration and Production Services in Houston, Texas. It was a small research laboratory that studied improvements that could be made in oil production out in the field. It was a great job, and I was surrounded by a lot of recent graduates who had been hired by Texaco.

Daddy Ben and I communicated a lot by telephone and in writing in the 1970s. On one of my trips in 1979 while I lived in Texas, Mom joined me in Dallas to celebrate the sixty-fourth wedding anniversary of her parents. I invited Daddy Ben, and he visited us and my grandparents in Dallas. A picture from that visit is shown in Figure 15-2. It is one of the few pictures I have of the three of us. I was almost twenty-six years old, Mom was almost fifty-five, and Daddy Ben was sixty. That was a very happy day for me, and I felt complete. My father was a sweet man who brought an anniversary gift to my grandparents, and we all had a nice time catching up with each other.

Figure 15-2: Daddy Ben, Jan, and Jan's mother (Dallas, Texas, 1979)

In the mid-1970s, Daddy Ben moved into a larger house in Fort Worth, as his children were growing and had many activities. He was able to increase the size of his woodworking shop and continued to make many kinds of coach and wagon models from scratch. He was a man of many talents, and now he was able to pursue those things of which he had always dreamed. He even picked up a new skill—sewing! If he had any spare time, he did small jobs as a commercial artist as well. His business card is shown in Figure 15-3.

Figure 15-3: Ben Smotherman post-retirement business card

One of Ben's passions was western art, maps, and history. I found the beginnings of a book that he was writing about western history, complete with detailed maps and a chronology of events in the expansion of the West (Figure 15-4). From the days of his Bud Tuttle stories, he had come a long way with his writing, but the topic had not changed very much!

While living in Houston, I learned a lot about geology and petroleum engineering, and I enjoyed my job at Texaco very much. I also was able to go to JSC in 1978 and view the Space Shuttle as it sat atop a Boeing 747 aircraft at Ellington Field. This version of the Space Shuttle (dubbed the Enterprise) was being used across the country for testing, and it was in Houston during a stopover. It was a thrill to see the new space vehicle, and I was hopeful that one day I would be able to work in the space program.

I tell young people that the path to a goal is often not a straight one. Not only did I change majors at Georgia Tech, but I also went to another university to obtain a degree so I could get a job. Then, I was not able to find a job in the space industry but took the highest paid job I could, which was in the oil industry. From every one of those experiences, I tried to make the best of the situation and learn from everything and everyone around me. They all helped me to become the person that I am, and the broad experiences and education I received from that circuitous path contributed to the skills that I would need as an astronaut.

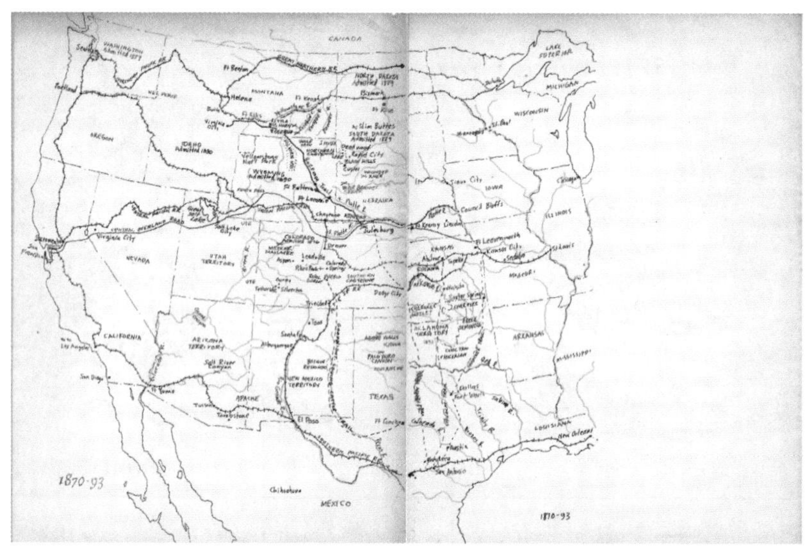

Figure 15-4: Ben Smotherman hand-drawn map of western United States, railways, roads, towns, Indian tribes, trails, and other features for period 1870 – 1893

In the late seventies, NASA was gearing up to fly the Space Shuttle. For the first time in about ten years, they were hiring again. So, we moved back to Huntsville in the summer of 1979, and I was ecstatic to have an interview with NASA in my hometown!

There was a lot of anticipation and excitement about the development of a reusable spacecraft, known as the Space Shuttle. On the Shuttle, experiments could be performed, extravehicular activities (EVAs) could occur, satellites could be deployed with a robotic arm, and classified missions could be conducted. The Shuttle would take off like a rocket and land like an airplane, or glider, on a runway.

To participate in missions on the Space Shuttle, a new type of astronaut was needed. No longer would military test pilots be the only members of a crew. The crew would consist of a pilot and commander, who were military test pilots, and mission specialists. The latter could be engineers, scientists, doctors, military personnel, or whoever met the technical qualifications.

A new cadre of applicants was opened when the astronaut selection criteria were shared in 1976. In January 1978, the eighth group of astronauts was announced, and it was the largest group ever selected. Numbering thirty-five people, the new astronaut class named themselves "Thirty-Five New Guys" or "TFNG."

I remember this astronaut selection very well because it included not only service members but also nonmilitary astronauts, women, and people of color. There were fifteen pilot astronauts in the group and twenty mission specialist astronauts. Included in the class were six women, three African American men, and one Asian American man. This selection opened doors to others like me and provided inspiration in that anyone who could meet the technical and medical requirements could become an astronaut.

When I interviewed for a job as an engineer at NASA, I could not believe that I was in the same place where von Braun and his team had worked when MSFC was formed in 1960. As part of my interview, we drove around the test stands where the Saturn V engines had been tested during my formative years in Huntsville. It was the first time that I had ever seen them up close—they were massive and impressive!

The place was abuzz about the new Space Shuttle program, and MSFC was responsible for the propulsion for the Shuttle. The Space Shuttle (or Space Transportation System) consisted of four main elements, as depicted in Figure 15-5. It carried up to eight astronauts and up to fifty thousand pounds of payloads (experiments, telescope, module, etc.) into low Earth orbit. The orbiter was the airplane-like vehicle that carried the crew and the payloads. The propulsion elements for the orbiter were the three Space Shuttle main engines (SSMEs).The external tank (ET) was a big fuel tank strapped onto the bottom of the orbiter that contained tanks of liquid hydrogen and liquid oxygen, which were fed to the SSME for its propulsion. The two solid rocket boosters (SRBs), including the reusable solid rocket motor (RSRM), were attached to each side of the ET and were fueled by a solid propellant. The SRBs provided over 80 percent of the thrust needed to get the Space Shuttle off the pad, each providing three million pounds of the designed 7 million pounds of thrust.

Figure 15-5: Space Shuttle elements (NASA photo)

When its mission was complete, the orbiter would reenter the Earth's atmosphere and land like a glider at either KSC in Florida or Edwards AFB in California. MSFC was responsible for managing and developing propulsion elements—the ET, the SSME, the SRBs, and the RSRMs. JSC in Houston was responsible for the orbiter and for engineering integration of the entire vehicle. The assembly of the vehicle in the Vehicle Assembly Building and the launch of the Shuttle from Earth were accomplished at KSC.

The timing was right for me to go to work in the space program. In fact, I firmly believe it was God's timing because NASA had just received authority to hire some young engineers to help with the Space Shuttle program and other major programs that would use the Shuttle. I was "over the moon" when I found out that I had been hired and would be working in the structural engineering area, which was my first choice.

On October 1, 1979, I took my oath of office as a United States government civil servant for NASA MSFC in the same building where so many previous MSFC employees, including Wernher von Braun and his team, began their NASA careers. It was my dream, and with the downturn of the space program in the 1970s, I had not thought it would ever be possible.

My first stop on my first day at NASA was to go to the office of my director, Alex McCool, on the top floor of Building 4610. He took me to meet my new branch chief, Mr. Paul Frederick, and the men (yes, all men) in the branch where I would work. Mr. Frederick was an incredibly nice man with a quiet voice, a steadying influence, and a great technical ability. Everyone else was friendly and knowledgeable as well. However, I felt that I had to prove I could do the job as a woman and a young person.

I was assigned to work on the structural analysis of the Space Telescope (ST), which would later be named the Hubble Space Telescope (HST). The ST was a program that was being managed by MSFC, and it was going to be deployed by the Space Shuttle. Goddard Space Flight Center was responsible for the development of the scientific instruments on the telescope. It was massive, weighing 25,000 pounds. This telescope was going to be able to see ten times further into space than any other space-based or land-based telescope and therefore would unlock discoveries about the universe that had never been possible before. Above the Earth's atmosphere, it was assured of clear observations free from light dispersion by air, light pollution, and contamination. It was unbelievable to me that I would play a small part in the development of this national asset.

The ST consisted of two major parts and a series of scientific instruments that could be changed out. The "guts" of the telescope were located in the Optical Telescope Assembly, which held the almost eight-foot diameter primary mirror and a smaller secondary mirror as shown in Figure 15-6.

I was assigned to work on the graphite epoxy structure metering truss that would hold the primary and secondary mirrors. The job of those in our organization was to make sure that the structures were strong enough to withstand the multitude of vibration and external loads that would occur during a Space Shuttle launch. We had to make sure that those huge pieces of glass in the telescope would not break and that they would maintain their alignment relative to each other. Also, it was a challenge to determine the variable material properties of the graphite epoxy, which was a unique, handmade material that ensured the telescope metering truss kept its dimensions constant in the harsh space environment.

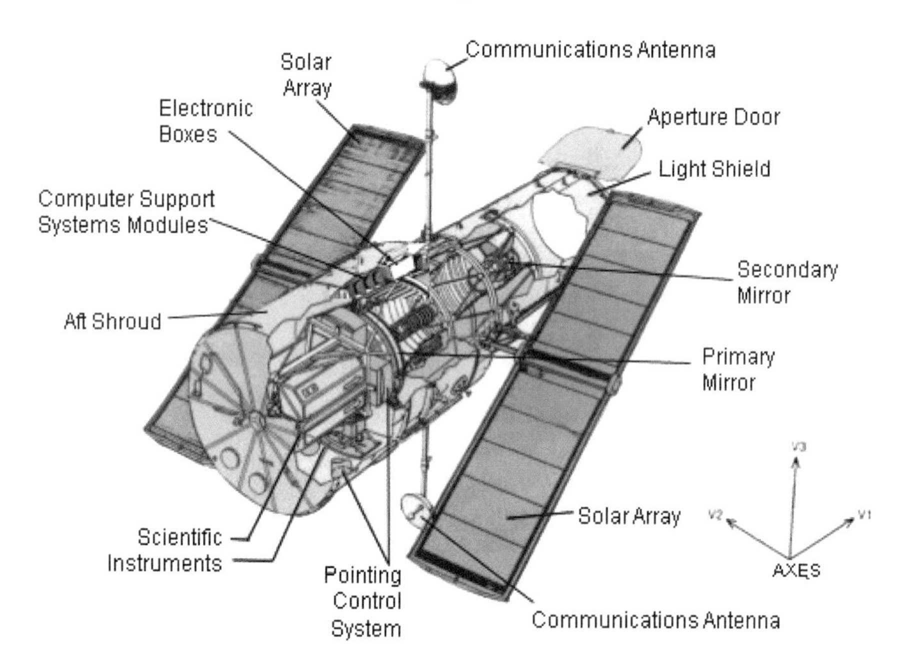

Figure 15-6: Cutaway diagram of Hubble Space Telescope, showing Optical Telescope Assembly inside the outer shroud (NASA diagram)

Being back in Huntsville was wonderful, and I began to ice skate again. As I was not jumping and spinning as much, I honed my ice dance abilities and continued to pass progressively more difficult figure skating and dance tests. I was also able to ice skate with Mom and cherished our time together.

It was pure joy to take advantage of the outdoors in the North Alabama region again as well as other places in Alabama and Tennessee. I discovered a network of outstanding and beautiful state parks, and it was a welcome relief to be able to commune with nature after being at work.

During this period, I knew that I wanted to continue my studies in engineering, so I enrolled at UAH to work on my master's degree in mechanical engineering. I started graduate school in January 1980 and was fortunate to have classes at night and for NASA to pay my tuition. I had the full support of my friends, coworkers, and supervisors at NASA.

One of my ice skating friends, an older gentleman who was a speed skater, was going (slowly) around the ice rink with me one night shortly after the 1980 astronaut group was selected. He asked me, "Why don't you think about becoming an astronaut?" Honestly, I was taken aback because that thought had never crossed my mind. He said, "You should think about it—you have an engineering degree, they are accepting women, and you work for NASA." It was then and there that I started considering becoming an astronaut.

To start, I became scuba certified so that I would qualify to dive in the Neutral Buoyancy Simulator (NBS). The NBS was used to simulate EVAs because floating in water was the best way to simulate working in space on a "spacewalk." Whenever tests or training were being done with the astronauts, it required a bevy of support divers to keep the astronauts safe, to assist with supplying tools and other equipment as needed, and to provide video or still photography. As I was an engineer on the ST and because my supervisors knew that I wanted to be an astronaut, they allowed me to be a support diver whenever a test or training was being done in the NBS with the ST. This was a meaningful experience for me, and I met many astronauts, including Jerry Ross, Bruce McCandless, Kathy Sullivan, Pinky Nelson, and Tony England, among others, while I was diving in the NBS. They were all very nice and helpful, but Jerry Ross in particular spent a lot of time with me and answered my many questions about how to become an astronaut.

The year 1981 was historic for the Space Shuttle program. After many delays, problems with the tiles on the bottom of the orbiter, issues with the SSMEs, and other challenges, the first flight of the Space Shuttle occurred on April 12, 1981. Coincidentally, that date was exactly twenty years after the first human, Russian Yuri Gagarin, went into space. Known as Space Transportation System 1 (STS-1), the initial flight of the Space Shuttle only carried two crewmembers: veteran astronaut John Young and rookie astronaut Bob Crippen. It was an ambitious mission because this was the first time a human spaceflight had occurred in a vehicle without a test flight. This first flight lasted only fifty-five hours, and ultimately, the orbiter glided down the runway at Edwards AFB after a successful mission.

My increasingly responsible work on the ST occurred while I was in graduate school, and I started taking flying lessons as well. I had always wanted to be a private pilot, partly because my father was a pilot and partly because I thought it might help my chances of becoming an astronaut. Most of all, I just wanted to learn to fly. In fact, I'd wanted to learn to fly as long as I could remember. I just had to wait until I could afford it!

On June 5, 1982, I began taking flying lessons in a Cessna 150 airplane at Moontown Airport (3M5), a grass strip 2,180 feet long and very wide. There was a hill on one side of the airport and another hill at one end of the runway. It was a challenging field from which to fly, but I felt it would be a good way to learn. It was also a fun place to fly, as there were many old and even homebuilt airplanes there.

Flying is one of the most enjoyable activities I have ever done. The sensation of being in the air and watching the world go by is not like any other experience I've had. Safely piloting the aircraft took all my attention, and I could not think about anything else. Luckily, I could block out the rest of the world by concentrating on flying and experiencing the beauty of seeing the world from the air.

I took my solo flight on August 24, 1982, and the rush of flying completely by myself is something I will never forget. I was so thrilled. After my solo flight, I received training on navigation, flew my solo cross-country flights, and started night flying.

Before long, I finished my flying requirements and ground school and did well on the written private pilot test. It was a happy and momentous occasion when I took my flight test and passed on June 9, 1983. I was now rated as a private pilot for single-engine land!

My father and I did not communicate as much once I was married and busy working. During that period, we usually only wrote on birthdays and holidays. However, I resumed writing to Daddy Ben in the early 1980s when I started flying. This really gave us something in common to discuss, and we did so frequently. It was during this time that Daddy Ben talked with me about his own flying career.

In June 1982, he wrote:

I am delighted that you are learning to fly and hope you keep it up. I wish we did live closer so we could talk flying. You can learn a lot that way. I'd like to help you avoid some of the mistakes I made in my early flying days. You are more mature than I was at the time and won't be inclined to 'show off.' After nearly killing myself a couple of times, I decided that it didn't pay to 'show off' in an airplane. There is a saying in the Air Force and among flyers: 'There are old pilots, and there are bold pilots, but there are no old, bold pilots.' I had many close calls but not from showing off. My advice to any young person, whether riding a bicycle, driving a car, or flying an airplane, is this: know your limitations and the limitations of the vehicle you are operating.

This was sound advice, indeed, and I always remembered it.

He continued:

Years after I quit showing off, I witnessed the broken, dead bodies of two pilots in a little Luscombe—an all-metal light plane. They were buzzing the beach near PAFB, showing off, and decided to show the bathers a slow roll. That airplane was not designed for aerobatics, and they didn't make it. The wreckage was so crumpled you could have loaded it into a pickup. My last attempt at showing off was trying a snap roll in a forty-horsepower Taylorcraft with a two-hundred-pound passenger. I fell out in a stall but fortunately had enough altitude to recover—just barely. You won't be tempted to try things like that. If any of your young flight instructors try it, tell them I said to get you back on the ground immediately—and don't fly with them again.

Daddy Ben even gave his E6B flight computer to me, and I used it as my own flight computer. He wrote:

It's a flight computer (circular slide rule type) that I carried in my flight suit. It was well used! I also used another one with more functions on it (to compute drift angle, etc.), but this is the one I was never without and the one I actually carried with me. When you start flying cross country, it may still be useful to you! But with today's navigation systems, distance measuring equipment, etc., not many professional pilots carry pocket computers anymore.

Also in this package was one of the Bibles my mother gave to him and the diary that he wrote to my mother when he was at war. I carried that Bible with me whenever I piloted an aircraft and was able to take it on one of my spaceflights. Daddy Ben told me that he was proud to have me carrying on the flying where he left off.

After I earned my pilot's license and was able to carry passengers, I flew frequently and carried family members and friends. I also took cross-country flights to places I wanted to go and enjoyed flying immensely.

When I sent a copy of my pilot's license to my father, he told me:

Whether you realize it or not, it puts you in still another special class. You will come to realize that more and more as you acquire your flying hours. For one thing, the lives of your passengers will depend on your knowledge and skill, and the fact that you accept that responsibility places you in a very special class of people. Safety of the ship, safety of the passengers, safety of yourself—those are the priorities you willingly accept, in that order, and it makes you a very special person.

That was sage advice, and I am so glad I had my father, an outstanding pilot, sharing his wisdom with me.

In 1983, two other flying friends of mine and I purchased a Cessna 172 (Figure 15-7). It was a nice airplane and was fully equipped for flying on instruments. Therefore, I started taking instrument flight lessons because I felt that I could be a safer and better pilot if I were instrument rated. Daddy Ben explained to me how he had flown on instruments back in the day and even included drawings! In a letter he wrote in December 1982, he said: "I'm a little sorry you missed what I call the romance of the early flying, but at the same time, I'm glad you have higher technology to help you along."

Another piece of good flying advice my father gave to me was: "The airways weren't too populated in those [his flying] days. The heavy traffic today is the greatest hazard which faces you. Keep your head on a swivel!"

It was great to talk about flying with my father, and I was so glad to get advice from a veteran who had thousands of hours of flying time. It brought us closer together, and that was one of the main benefits of my learning to fly.

Figure 15-7: Jan at the controls of her Cessna 172 airplane

In the meantime, I continued working on my master's degree at UAH and did some research about filament wound composites for my thesis, which was titled, "The Time Dependent Response of Filamentary Composite Spherical Pressure Vessels." In May 1983, I earned my Master of Science degree in mechanical engineering. By that time, it was well known among my friends, family, and coworkers that I was going to apply to become an astronaut one day.

In 1983, an announcement came out that there would be another astronaut selection in 1984 for the Space Shuttle program. So, on my thirtieth birthday, I sent my astronaut application to JSC. At that point, all I could do was wait to see what would happen next.

After completing my master's degree, I knew that if I was serious about becoming an astronaut, I needed to continue my graduate education and earn a Doctor of Philosophy (Ph.D.) degree. I became aware of an outstanding program that the government offered to its civil servants: the full-time study program. It was a competitive program, and only a few people were selected each year to participate.

NASA's full-time study program gave students a year off work to study full time while their NASA salaries were fully paid. When I applied for this program, I emphasized that I wanted to be a NASA astronaut, and even if I was not selected, I planned to stay employed with NASA because I was passionate about the space program. I was sincere and honest in my application, and in the summer of 1983, I was selected to enroll in the full-time study program at NASA.

I started my coursework toward my Ph.D. in the fall of 1983, and it was so nice to be able to take a full course load since I was a full-time student! I enjoyed the courses that I took, most of my professors, my fellow graduate students, and liked having an office at the university. My research was a more in-depth study into time-dependent characteristics of filament wound composite materials and involved much more complicated geometry and mathematics than for my master's degree. However, I knew that I needed that "ticket" if I was ever going to be an astronaut.

Chapter Sixteen
Beyond the Sky

My child so dear, with smile so bright
Your dreams to try, "I want to fly
Into the sky, into the sky."
Her thoughtful ways, her swinging hair
And azure eye, "I want to fly
Into the sky, into the sky."
Love and prayers, acceptance reigns
With eagles high, "I want to fly
Into the sky, into the sky."
This child so dear, who soars and climbs
Into misty sky, "I want to fly
Beyond the sky, beyond the sky."

To Jan, with love, Mom 1-8-88
(Dolly Jo Gantz Smotherman Davis)

As I mentioned previously, Space Shuttle astronauts were divided into two categories: pilot and mission specialist. There were different medical and technical requirements for each type of astronaut.

The pilots were typically graduates of the military test pilot schools and were required to have a certain number of flying hours in high-performance aircraft. The Space Shuttle commander and pilot were always pilot astronauts and were responsible for flying the Shuttle during all phases of flight.

The mission specialists were technical specialists in the fields of engineering, science, medicine, etc., and were responsible for carrying out the mission of a Space Shuttle. They were in charge of conducting the EVA, operating the Canadian mechanical arm (CANADARM), conducting scientific experiments, acting as a "flight engineer" for the pilot and commander, and doing whatever was necessary to carry out a successful mission. As I told people, "The mission specialists get to do all of the fun stuff!" However, I'm sure my pilot astronaut colleagues would say that they did fun stuff too.

In 1984, almost five thousand people applied to be astronauts, and approximately one thousand of them did not meet the minimum criteria. The next step in the process was for the selection board to determine who was "highly qualified" in each category and to get reference submittals for those. It was encouraging to me that the people whom I had listed as references on my application were getting requests from the astronaut selection committee!

In those days, the next step in the selection process was a week in Houston for extensive medical testing and examination and an interview with the astronaut selection board. On the evening of Friday, March 9, 1984, I received a call from the astronaut selection office at JSC. When asked if I was interested in coming to Houston for an interview, I just couldn't believe it! Of course, I was!

Each week for six weeks, about twenty astronaut interviewees headed to Houston. The makeup of those twenty was approximately the same as a typical astronaut class—about half mission specialists and half pilots with a mixture of engineers, scientists, people from NASA, and people from academia as well as a combination of genders and ethnicities. Our group was the fifth week of interviewees, and there was only one week of interviews after ours.

I received my travel orders from NASA and flew to Houston on Sunday, March 11. Our first briefing was that evening at the hotel, and it was given by astronaut John Young! John was head of the astronaut selection board and, as I've mentioned, had been on the first Space Shuttle mission. He also had flown on a Gemini mission and was one of the men who had walked on the moon. As a Georgia Tech graduate, he was one of my personal heroes. I kept saying to myself, "That's John Young! That's John Young!" It was the beginning of a fun and eventful week.

Our interview group consisted of twenty-two people from varied backgrounds and ethnicities and is shown in Figure 16-1. There were Army, Air Force, and Navy candidates and people from NASA, academia, and industry. The group was about half military and half civilian. All were very sharp and accomplished individuals.

Figure 16-1: Jan's astronaut candidate interview group in 1984; eventually, six of these candidates would fly into space

We were encouraged to meet as many astronauts as possible and were free to roam around Building 4 where the Astronaut Office was located. The first person I wanted to talk with was Jerry Ross, with whom I had spoken about being an astronaut and who worked on HST. He was also a mechanical engineer, so we had similar backgrounds. Jerry was as friendly and approachable as ever, and he outlined what the week ahead was going to look like.

The week consisted of tours of the training facilities the astronauts used as well as medical exams, medical exams, and medical exams. I was like a giddy child on the tours, as we were given a behind-the-scenes look at mission control, the mockups of the Space Shuttle in Building 9, the motion-based simulator and fixed-base simulator in Building 5, the

T-38s and aircraft operations at Ellington Field, the Weightless Environment Test Facility (WETF) where astronauts trained underwater for their EVA, and so much more.

For a thirty-year-old, the medical examinations included tests I had never had before, like a proctology and an electrocardiogram. There were many tests on the eyes, including a flash picture of the retina—ouch! Of course, we had the requisite hearing tests, pulmonary function test, bloodwork, urinalysis, X-rays, heart stress test (treadmill), motion sickness test, musculoskeletal exam, etc.

It was no surprise that there were also psychological tests and interviews. To check for claustrophobia, I was zipped into a three-foot diameter ball (with breathing air, heart monitors, and a microphone) and asked to evaluate the ball (known as the personal rescue sphere) as a potential rescue method for a stranded astronaut. I found the interior of the ball a very pleasant and relaxing environment. It was dark with white noise, and I was sitting in almost a meditative position.

There were two opportunities to interact with the selection board. One was the actual interview, which was one to two hours, and the other was a social with barbeque and beer at a famous astronaut hangout, Pe-te's. The interview was the most stressful time of the week because it was a critical part of the selection process. From talking with the other candidates who were interviewed before me, I knew what some of the typical questions were. In addition to telling my life story, I knew that I would be expected to share what musical instruments I played and in what sports I participated.

The purpose of the social at a good old Texas barbeque was for the board to see how well the interviewee could handle social events, mix and mingle, and practice table manners while imbibing beer. I was perfectly comfortable in a social setting and met as many people as I could, "working the crowd."

The twenty-two of us became fast friends that week and had a great time eating out together and sharing information. It was not a competitive environment at all; the competition had been getting to the interview. After that, everything was up to the individual and was not dependent on others. Duane Ross, the head of the astronaut selection office, stated,

"They [interviewees] should follow their own passion and aptitude, and their character and accomplishments will speak for themselves."

There were 4,934 applicants for astronaut in 1984, and 128 were interviewed. I felt very fortunate and lucky to be interviewed that year, but I was not selected. I was disappointed, to be sure, but I knew that at least I had a chance and perhaps could be selected later. For a thirty-year-old, I had the best physical examinations that I could imagine, and I was comforted to know that I'd passed all of the medical requirements for astronaut mission specialist.

On May 25, 1984, seventeen people making up the tenth class of astronauts were selected. This was the third group of Space Shuttle astronauts, who called themselves "The Maggots." Seven were pilot astronaut candidates, and ten were mission specialist astronauts. I was delighted that three people from my interview group were selected: G. David Low, Mike McCulley, and Kathy Thornton.

The chance of becoming an astronaut then seemed attainable. One of my favorite quotes is, "We may not always reach our goal, but there is recompense in trying. Horizons broaden so much more the higher we are flying (Anonymous)." By being interviewed to become an astronaut, going on tours at JSC, and meeting many more astronauts, my horizons had broadened greatly. I did not reach my goal, but I was more determined than ever to finish my Ph.D. degree.

I returned to Huntsville, Alabama, with new inspiration and motivation after my whirlwind week in Houston. I had a renewed purpose in buckling down on my studies. For the remainder of 1984, I didn't do much except study, work on my dissertation, and finish my coursework.

In addition, I had to take my qualifying examination and defend my dissertation. I was not too worried about the latter, but I was really concerned about the former. My entire committee would administer the oral exam and ask me anything they wanted. I was reminded that I was getting a Doctor of Philosophy, not a Doctor of Engineering degree. Therefore, I needed a broad and deep knowledge of many subjects. Furthermore, I could not use any aids, references, or notes.

I studied feverishly for this test because everything I was working on and my potential future career as an astronaut was dependent on it.

Unfortunately, I did not pass this qualification exam for my degree. Of course, I was devastated because I had studied so hard, and I knew I had to pass to achieve my goals. When I discovered the news, I was reduced to tears, which was embarrassing, but I was very upset. At that point, I buckled down further, determined to pass on my next attempt.

The next time I took the qualifying exam, I thought the questions were fairer than the first time, and I did a better job answering them. I entered the room with confidence I would not stumble. And to my relief, I passed!

The whole ordeal of the qualifying exam took me to my limits emotionally, physically, and intellectually. However, it made me stronger because I would not quit, and I worked hard to make sure I would succeed. When I give talks to young people, I tell them this story and talk about how perseverance and hard work overcame a tough situation. I also tell them to keep trying and not give up!

Writing the dissertation was a challenge, as my advisor had endless rewrites for me to do, and it seemed like it would never end. However, it did eventually end, and I had no difficulty defending my dissertation, putting me on the road to graduate in June 1985.

Figure 16-2: Jan receiving her Ph.D. diploma from the University of Alabama in Huntsville, 1985

I received my diploma from Dr. John Wright, president of UAH (Figure 16-2). It remains one of my proudest accomplishments, as I not only learned about the time-dependent loss of strength and viscoelasticity of spherical filament wound pressure vessels, but I also learned a lot about myself.

In 1985, it was announced that there would be another astronaut selection, but the pool of candidates would be from the same pool of candidates used in 1984. Therefore, a new application was not necessary unless information needed to be updated. Of course, I added the upcoming completion of my Ph.D. to my application and sent the update to the astronaut selection office.

NASA considered thirty-three civilians from the selection roster developed during the 1984 selection process and 133 nominees from the military services. Only fifty-nine of the highest-ranking applicants were interviewed and given medical evaluations at JSC. I was sure that I would be interviewed this time since I had just the "ticket" I thought they were looking for: the Ph.D. However, that was not the case. The eleventh class of astronauts (and the fourth class of Space Shuttle astronauts) consisted of thirteen people and was announced in June of 1985.

After that, I returned to my work as a structural engineer at MSFC. There was still a lot of work to be done on the HST, and I was assigned to work on the solar arrays. This was a very interesting project, as the solar arrays were unfurled with an ingenious method, and there were challenges to be met.

The HST solar arrays were provided by the European Space Agency (ESA), which was based in Noordwijk, the Netherlands. Therefore, in addition to working on an unusual engineering challenge, I was able to go to the Netherlands, which was my first trip to Europe. I also enjoyed working with the engineers at ESA, although in the mid-1980s, they weren't used to collaborating with very many women! I am pictured beside a model of the HST in Figure 16-3.

Another project to which I was assigned after I returned to work full time at MSFC was the Advanced X-Ray Astrophysics Facility (AXAF). This was another one of NASA's Great Observatories, designed to observe the cosmos in the X-ray region. Due to the short length and high energy of X-rays, the telescope design was vastly different from that of the Hubble (which was an optical telescope). The AXAF set of mirrors were nested conically shaped pieces of glass, so the X-rays could glance off of the glass in the telescope and into the scientific instruments. It was fascinating science, and I really enjoyed learning about a new facet of astronomy.

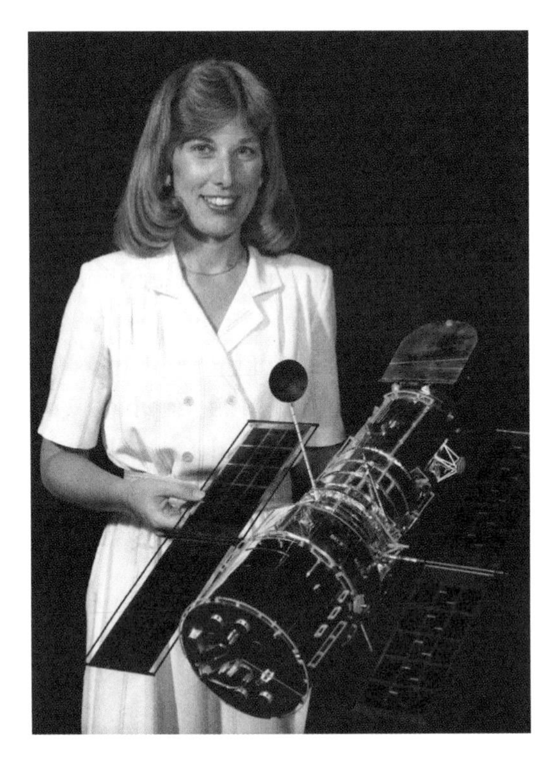

Figure 16-3: Jan and a model of the HST, 1986 (NASA photo)

My work on the HST allowed me to still be involved as a scuba diver in the testing of the telescope in MSFC's NBS. During one of those tests, photographers from *LIFE* magazine took some pictures, which featured the assigned EVA flight crew for the HST, Kathy Sullivan and Bruce McCandless. *LIFE* was going to publish an article about the great missions occurring in 1986, including the HST and the Shuttle flight of teacher Christa McAuliffe. The photograph they took that day for the magazine included me as one of the scuba divers and is shown in Figure 16-4. The cover article, titled, "Seeing Beyond the Stars: A Preview of America's Biggest Year in Space" was published in the December 1985 issue of *LIFE* magazine.

Figure 16-4: Jan (bottom center) as a diver in the Neutral Buoyancy Simulator at MSFC with HST astronauts Bruce McCandless and Kathy Sullivan (photo courtesy of Life Magazine, December 1985)

Tragically, although 1986 started as a year of promise and hope in the space program, it was not to be. On January 28, 1986, the launch of the Space Shuttle Challenger with teacher McAuliffe as well as crew Dick Scobee, Mike Smith, Judy Resnik, El Onizuka, and Greg Jarvis ended with an explosion after only seventy-three seconds of flight.

At work at NASA MSFC, we all usually crowded into a conference room to watch a Space Shuttle launch, and the launch of Challenger mission 51-L was no exception. I will never forget the gasps and tears from those witnessing the disaster unfold. We left the conference room without saying a word. Immediately, we all were assigned to comb through data of the Space Shuttle propulsion elements for which MSFC was responsible. I was in shock and couldn't believe that Judy Resnik had been one of those on board. She was a woman I admired whom I'd met during my week in Houston for astronaut interviews.

My mom called me later in the day to see if I was okay, and she asked me if I still wanted to be an astronaut. "Absolutely. This doesn't change my mind—I want to fly" I answered resolutely.

It was quickly discovered that a puff of black smoke had come out of a joint of the SRB. Video footage revealed that the black smoke and resulting torch of flame from the SRB had melted the strut holding the SRB to the ET. This caused the SRB to rotate into the ET. The ET exploded, and subsequently, the vehicle exploded.

President Reagan formed the Rogers Commission to investigate the accident, and their report was released on June 9, 1986. The technical cause of the accident was found to be a failure in the O-ring's sealing of the aft field joint on the right SRB. Each joint where SRB sections had been put together at KSC was known as a "field joint." The Rogers Commission criticized NASA for having an inadequate safety organization and insufficient communication between its managers.

I continued to work on the HST and AXAF programs throughout 1986. At this point, I thought that my chances at becoming an astronaut were dashed, as I was from MSFC, which was being blamed for the Challenger disaster.

Meanwhile, many astronauts were leaving the astronaut corps in 1986, fearing long waits and uncertainty about the future. Therefore, NASA put out another call for astronauts to apply for a new class to be selected in 1987. Again, I submitted my application, this time with little hope that I would be interviewed.

The new MSFC center director, J.R. Thompson, and new management in the Space Shuttle offices as well as a new organization at NASA for safety, reliability, and quality assurance carried on the work as best they could. There was a focus on responding to the Rogers Commission's recommendations and reporting to the president about how NASA was going to proceed with the Space Shuttle program.

Late in 1986, I received a call from Thompson because he wanted to give me a new assignment: to be the lead engineer for the redesign of the SRB aft external tank attach (ETA) ring. This ring attached the SRB to the ET and was about fourteen inches away from the field joint with the O-ring that had failed during the Challenger launch. The aft ETA ring was a C-shaped ring around the booster, and some of the bolts at the end of the "C" were failing. I had my work cut out for me and was immediately busy. I was no longer able to work on HST or AXAF, but I was glad to be focusing on something that would get the Space Shuttle flying again.

The holidays and the beginning of 1987 were a blur because I was totally consumed with helping to determine the structural characteristics of the aft ETA ring and come up with a solution. In addition to a demanding regimen at work, my personal life was not going well, and I separated from my husband. We divorced later that year.

The new assignment of redesigning the SRB aft ETA ring was a challenge and one which was heavily reviewed. Fortunately, my team had the time and ability to do the technical work, and I was busy "fighting fires," making presentations, and answering questions from people internal and external to NASA.

A positive and uplifting event amid my tough personal life and pressure-packed work life was a telephone call in mid-February from JSC to ask me to come to Houston for an interview for the astronaut class of 1987. I was surprised but very happy! I was in the first week of interviewees, which was a good sign.

When I went to Houston for the week of interviews and medical examinations starting February 23, 1987, I was more relaxed because I knew what to expect. The interview group numbered twenty-two people and included someone with whom I'd interviewed in 1984—a Navy pilot named Bill Readdy who was then flying in the Aircraft Operations Division for JSC.

John Young was once again on the astronaut selection board. He had been very outspoken about the Challenger accident and offered his own ideas about factors that could have caused the tragedy. About one week before I went to Houston, an article had been published in *Aviation Week and Space Technology* magazine about one of John's theories. He thought that the failure of the O-ring that had caused the accident in the SRB field joint might have been due to deflection of the SRB aft ETA ring.

During the interview, he asked my opinion about that theory. I told him that I did not think the aft ETA ring could have contributed to the accident. In addition, I had data (in my head) and an explanation that I'd given to others who'd asked me the same question. During my interview, the dialogue between John and me lasted several minutes, and I was honest with my opinion. I admit that it was intimidating to disagree with John Young, but I had the facts and the supporting information, so it was not a subjective opinion. I was not too concerned that disagreeing with John Young would hurt my chances of being selected as an astronaut because I was telling the truth, and that was important to me. Fortunately, I had read the article beforehand, so I was prepared to answer any questions John had.

After my interview, there was a lot of excitement and anticipation about being selected. I was very busy at work, so I focused on that and tried to not think too much about the astronaut selection. I believed my chances were slim to none, and I was diligently doing my part to get the Space Shuttle flying again.

After much design, analysis, and review, my SRB aft ETA ring team concluded that the best redesign to ensure a safe launch and return of the SRBs was to make the ring 360 degrees around the SRB. I felt good about this approach and was confident in presenting our solution to NASA's upper management and the various review panels.

During the summer of 1987, the American Society of Mechanical Engineers (ASME) had its annual meeting. In ASME, I had risen to be a member on a national committee; therefore, it was necessary for me to attend the meeting in June 1987 in Toronto, Canada. When I arrived there, my ASME colleagues were abuzz with the news that someone from JSC was trying to get in touch with me. My mother had also called and told me that George Abbey, the head of the Flight Crew Operations Directorate in Houston at JSC, was trying to reach me, and she'd told his office my whereabouts.

George Abbey! Everyone knew if you were selected to be an astronaut, George Abbey would call you. If you were not selected, another member of the selection board would call you. I remember Carolyn Huntoon had called me in 1984, and as soon as she called, I knew that I had not been selected.

It was very late by the time I arrived at my hotel, so after phoning my mom and getting George Abbey's contact information, I decided to return his call the next day. I asked one of my best friends, Sue Skemp (who was also at the meeting), to sit with me when I made the call because I knew what it was about.

The morning of June 5, 1987, I contacted George Abbey. I was trying to act calm and collected, but I was nervous, excited, and happy all at the same time. We chitchatted a while, and he even asked about the weather in Toronto! I was clutching Sue's hand, and she swears that she still has scars from my fingernails!

Eventually, George asked me, "Are you still interested in becoming an astronaut?" In response, I said, "Yes, sir, I am!"

I don't remember the conversation after that, but it had something to do with reporting date, announcement of the class, etc. I was told not to tell anyone about my selection until after the press release later in the day.

Well, I had to call my parents to share the exciting news, but I made them promise they would not tell anyone until the press release came out. Of course, my mom and stepfather were extremely excited and proud of me and promised that they would not say a word until later that day when I gave the go-ahead. As I've mentioned, my mom told me to "put God first, then your country, and go for it!" That was very wise advice that I took seriously.

I did not know who else was in my astronaut class until the press release came out. I was pleased to learn that seven of the people with whom I'd interviewed in February were in the 1987 astronaut class with me! There were fifteen people in our class—eight mission specialists and seven pilots, all pictured in Figure 16-5.

Astronaut Candidates—Selected June 1987
Top Row: James S. Voss, Kevin P. Chilton, Curtis L. Brown, Jr., Andrew M. Allen, C. Michael Foale
Middle Row: Jan D. Davis, Gregory J. Harbaugh, Mae C. Jemison, Kenneth D. Bowersox, Bruce E. Melnick
Bottom Row: Mario Runco, Jr., Donald R. McMonagle, Kenneth S. Reightler, Thomas D. Akers, William F. Readdy

Figure 16-5: Jan's astronaut class—the class of 1987 (NASA photo)

I was elated but had to hold in my excitement until that afternoon. I had to pinch myself because I just could not believe I had been selected! My thoughts were a swirl of emotions, plans, and hopes for the future. It was just the thing I needed at that time. It was definitely God's timing, as it was going to be the right challenge and the right change for me.

Once I could tell people the great news, I spent quite a bit of time making phone calls. I let Daddy Ben know, and he was ecstatic and wished me well. I called family, friends, my NASA supervisors and mentors, and my ASME colleagues who were with me in Toronto. We had a big celebration in the Toronto Tower that night. I will never forget that memorable occasion!

Meanwhile, in Huntsville, my parents were being interviewed by print and television media. Huntsville is a space town, and they were (and still are) very excited to have their own astronaut.

Some friends sent congratulatory flowers to me when I was in Toronto, and I made a corsage from some of the flowers. While on the airplane ride home, I was seated next to one of my ASME friends who was from Mobile, Alabama. When the female flight attendant saw my corsage, she asked, "What's the special occasion?" My friend quickly answered, "She was just selected to be an astronaut!" The flight attendant responded, "Well, I guess you can teach a monkey to do anything!"

My friend and I were quite taken aback by that remark, so I remember it well. It was my introduction to the attitude that not everyone was going to be supportive of me being an astronaut (or believe that I was an astronaut). I had to learn to deal with it in a non-reactive way.

At the time of my selection in 1987, my brother was working in Atlanta. My flight connections took me through there, and he greeted me with irises (because I had admired them in Holland) and told me he was continuing to my destination with me! It was such a nice surprise to see him and receive his thoughtful gift, and I was pleased to know that he was going to accompany me to Huntsville. One of my favorite photographs (Figure 16-6) is the picture of my parents greeting us when we arrived at the Huntsville airport that day. It was a happy occasion, and it was so meaningful to be able to share it with family.

Huntsville and MSFC were still reeling from the results of the Rogers Commission and the Challenger investigation when my astronaut selection was announced. So, Center Director J.R. Thompson used the opportunity of my astronaut selection as a time to celebrate. Therefore, there was a "Jan Dozier Day" on August 7, 1987, with a big celebration at MSFC. Hundreds of people came out to celebrate with me in one of the high bay buildings at Marshall.

Figure 16-6: Jan's stepfather, mother, and brother greeting her at the airport after news of her astronaut selection (Huntsville, Alabama, June 1987)

That night, the city celebrated my selection with an event at the U.S. Space and Rocket Center, home of Space Camp. Many dignitaries were there, including local and state politicians and Senator Al Gore from Tennessee. I was awarded many plaques and gifts from various organizations as well my Ph.D. hood from UAH. It was great to have that support from MSFC and from Huntsville, and that support has remained throughout my career.

Other smaller yet meaningful parties were held by my division at work, my aerobics class, my sorority, my flying friends, and my team who had worked so hard on the SRB redesign. My friends and coworkers also threw a party for me by the Tennessee River. I could not have asked for a better sendoff than I had in Huntsville, and I probably had the best goodbye parties of anyone in my astronaut class!

Chapter Seventeen
The ASCAN Years

I was given August 17 through August 21 to make the move to Houston, and I asked my mother to drive with me. Mom was thrilled that I was back in her home state of Texas! I was also able to show her my new house, and we spent the rest of the week eating out and shopping for things for my home.

Mom never talked about it, but I wonder if she was thinking about when she endured the many months of flight training that my father had away from home and how she had to say goodbye to him before he went overseas to war. While I was not going to war, I was still going to be separated from her for a long time and was training for a risky and hazardous job. However, she was always supportive and never let me know her apprehension about the dangers of my chosen profession. I am sure that she was just as supportive of my father during his flight training and eventual combat missions.

I know my father was anxious to begin his flight training back in 1942, and I was just as anxious and excited to begin my own training. I knew that it would be challenging, but that is what I found exhilarating about the whole experience.

Our astronaut class reported for work at JSC on Monday, August 24, and spent the next two days getting oriented. For the rest of the week, we learned about the systems of the T-38 jet that we would learn to fly. The Northrop T-38 Talon is a two-seat, twin-engine supersonic jet used as a trainer by the Air Force and NASA. We were also issued a NASA flight suit, jacket, and name tag and ordered custom boots and helmet.

Almost every night that week, our class and spouses got together to go out to eat or do something socially. It was a fantastic group of folks. I remember my mother telling me that the friendships the Air Force families had were quickly formed and were bonding for life. I know that the friendships my father had with his fellow fliers and POWS were very close and lasted a lifetime. Now, I was experiencing that for myself. The first week of ASCAN (short for astronaut candidates) training was the beginning of an incredible experience, and I couldn't believe that it was actually happening!

Figure 17-1: In wilderness survival training at Fairchild AFB, making a signal out of parachute remnants (NASA photo)

Before we could fly in the T-38 jets, we needed quite a bit of training to ensure that if we had to bail out of the aircraft, we would know how to handle not only ejection procedures but also survival for wherever we might land. Therefore, the week of August 31 through September 3, we all went to Fairchild AFB in the state of Washington for our land survival training. If we ejected out of an aircraft in a remote wilderness area, we could be stranded for days, so we needed to know how to use the devices in our parachutes and seat kit to get help. We learned how to shoot flare guns, use a mirror for signaling someone, build a shelter with our parachute, operate the radio from our parachute harness, and use items from our seat kit to survive. In Figure 17-1, I am shown constructing a signal out of the remnants of my parachute that could hopefully be seen by a search and rescue team. We did not go hungry, as some had brought contraband to eat! It was great fun, and I learned a lot.

Figure 17-2: Ejection seat training at Vance AFB (NASA photo)

Figure 17-3: Jan learning how to use her parachute harness to steer the parachute at Vance AFB (NASA photo)

The parachute training my class received was at Vance AFB in Enid, Oklahoma, on September 10 and 11. This training also included ejection seat training (Figure 17-2) and instruction on the ins and outs of the parachute and parachute harness. After we practiced on a stand where we were hanging by a parachute harness (Figure 17-3), we were pulled up into the air by a truck and then had to land using the proper and safe position for a parachute landing fall. This was all new to me, but I did it, as shown in Figure 17-4!

The following week, those of us who were back-seaters (non-military trained pilots) spent the week in ground school and in preliminary flying with the T-38 instructors. I was glad that I'd had ground school already, even though it was for a propellor aircraft, as I could more easily absorb and understand what was being taught.

Figure 17-4: Jan taking flight in her parachute at Vance AFB (NASA photo)

The last wicket that we had to pass to fly the T-38s was water survival training, which we did at Homestead AFB in Florida from September 21 to September 23. Once again, we learned the intricacies of the parachute harness and how to release the parachute as soon as we hit the water. Our first exercise was to go from a tower down a long wire (which was at an angle to the water) until we reached the water. Then, we released our parachute; otherwise, the wind could catch in the parachute and drag us along or push us underwater.

To accomplish the next phase of training, there was a tower suspended over the end of a boat (Figure 17-5), and we were dropped into and dragged through the water. We had to then flip over so that we were on our back and could release our parachute.

The last phase of training was to know how to inflate and use the personal raft that was in our seat kit and then be picked up by a helicopter. We also had to know how to utilize a large raft that could be dropped down to us via helicopter.

Keep in mind that we were in a large class with Air Force lieutenants who were being prepared to fly aircraft for the service. The military pilots in my astronaut class were seasoned pilots who had been through this training many times. Although they took it seriously, they also had fun with it! Some of them decided to build a pyramid in their large raft (Figure 17-6). It was such a hoot, and the young airmen watched with amazement that we were able to get away with such shenanigans!

Figure 17-5: Jan learning water survival with her parachute harness and life preserver at Homestead AFB; she was dropped from a tower and dragged through the water by a ship (NASA photo)

Figure 17-6: Astronaut class of '87 having fun with the life raft during water survival training (NASA photo)

In our off time, our class had fun doing some scuba diving and snorkeling in the Florida Keys. In the end, I passed all the written examinations and practical tests that were required by the Air Force. I was cleared to fly in T-38s!

When looking at the procedures my father used for B-17s, I realized they didn't receive much training on how to bail out of a plane or how to utilize their parachutes on land or on the water. There were no ejection seats; instead, each crewmember was assigned an area of the plane where he could jump out. I was glad that I had so much training on parachute procedures and practice on how to use them.

On September 24 and September 25, my classmates and I received our initial T-38 flights. Mine was with NASA staff pilot Joe Tanner (who later became an astronaut). I have never had so much fun in my life! We did aerobatics, general flying maneuvers, and practice approaches. It was a much faster aircraft than I had flown before, so it took some quick thinking to be able to fly it. Back-seaters were not allowed to take off or land the aircraft, but I did just about everything else. That was the first time I traveled faster than the speed of sound, and that was a thrill! When the flight was over, Joe told me that he did not think he had ever been on a flight where someone enjoyed the flight as much as I did. A picture of me with Joe after the flight is shown in Figure 17-7.

Figure 17-7: Jan's first flight in a T-38, with instructor pilot Joe Tanner (who later became an astronaut)

Friday afternoons were generally reserved for flying for our class, and I just could not get enough. It was so exhilarating and fulfilling and much more enjoyable than I'd ever imagined was possible. Mission specialists had a requirement of four hours of flying time a month, which I had no trouble meeting!

For my first cross-country flight, I went to Huntsville, Alabama, to visit my parents. My brother was in town, so it was a momentous occasion. They were all so happy and proud to see me in my NASA flight suit, flying in a T-38 (Figure 17-8).

Figure 17-8: Jan's first trip home after being qualified to fly T-38s, with her family to greet her there.

We started our briefings/classes/tours at JSC the last week of September. In the Astronaut Office's large conference room on the third floor of Building 9, we were briefed on each Space Shuttle system by an astronaut, a Space Shuttle instructor, and an engineer or scientist responsible for the system. Wow—there was so much to learn! Each system was then taught to us in a one-on-one class in the Space Shuttle single systems trainer (SST), which simulated the systems, the failure signatures, and the operations of the system and taught us how to use the flight procedures, called flight data file (FDF).

I have a syllabus from this period, and the Space Shuttle systems that we studied were ascent; entry; guidance navigation and control (GN&C); aborts; caution and warning; communication; electrical power system; auxiliary power unit and hydraulics; orbital maneuvering system (OMS); reaction control system (RCS); main propulsion system; mechanical systems; Environmental Control and Life Support (ECLS) systems; displays; and controls, among others. We spent about a week on each system, which wasn't enough! If this is making your head spin, I understand because it made mine spin also. Fortunately, we were given many more chances to learn with textbooks, flight procedures, and more and more simulations.

We had briefings from nationally known experts on various fields of science or operations. The background and technical specialty of each astronaut candidate in my class was different, and the purpose of these briefings was to make sure we had all been briefed on the basic things that we would need to know as a Space Shuttle crewmember. For example, we had classes on life sciences (LS), avionics, materials science, geography, electronics, Space Shuttle elements, aerodynamics, computers, geology, loads and structures, NASA history, media relations, and on and on. It was fascinating! I like to discover new things, so I loved learning so much. These briefings continued through the end of December 1987.

During the next few months, we toured every NASA facility and center. There are ten NASA centers as well as other facilities such as White Sands Test Facility, Wallops Flight Facility, and Michoud Assembly Facility where the ET was being assembled. This was also a great opportunity to see the work NASA was doing around the country.

Another place we visited was New Mexico for a geology field trip with Bill Muehlberger, who trained the Apollo astronauts on how to find moon rocks. It is important to learn geology because geological formations studied from orbit are very interesting. It was a fascinating trip to a beautiful state, and we also experienced local culture and Indian folklore.

For lessons on astronomy, we went to a planetarium in Houston. We also had lectures from experts in their field, such as one on spacecraft design by the legendary Dr. Max Faget.

To achieve our T-38 flying time, we were allowed to fly all over the country. Flying to Huntsville was an easy hop, so I was able to come home to visit my parents many times. To me, flying T-38s was one of the best perks of being an astronaut. Not many civilian pilots have the opportunity to fly a high-performance jet.

In addition to visiting the NASA centers and spending time flying, the mission specialists started training for EVA and on the CANADARM in January 1988. Beginning with the Gemini and Apollo programs, the EVA training was done in a large pool at JSC. The astronaut trainee would put on the suit used for spacewalking with hookups for air and cooling water. The undergarment for EVA had tubes with water running through to keep the astronaut from overheating due to their own body heat. The divers who assisted in the simulation would put weights along the outside of the spacesuit so that the astronaut would not float up or sink in the water. Called neutral buoyancy, it was the closest thing on Earth to doing an EVA in space. Tools and mechanisms had floats on them to make them neutrally buoyant as well.

The astronauts have their own gymnasium at JSC because fitness and conditioning are paramount for flying in space and in high-performance aircraft. My favorite things to do to stay in shape were racquetball, bicycling, and walking in addition to working out with weights. I also enjoyed swimming and ice skating when I could find the time and the facilities.

In January 1988, Space Shuttle systems briefings continued with more detail and instruction on various aspects of GN&C and computers, mission rules, payloads and payload integration, crew escape, rendezvous, aborts, dumps/interconnects, landing/deceleration, ranging and tracking systems, flight planning, and so on. Science and other briefings also continued in January and included orbital mechanics, space station, crew duties, photography and Earth observations, astronomy, meteorology, oceanography, planetary science, space physics, and materials processing, among others. I was learning so much and was challenged to absorb an enormous amount of information. I enjoyed every bit of it!

By the first of the year in 1988, we were spending more time in the Space Shuttle simulators. The SST trained us on one system at a time, and each system took multiple classes so that we could fully understand it, its malfunctions, and its interactions with other systems. Gradually, we were introduced to the more sophisticated simulators where we put our knowledge of each Space Shuttle system to the test.

The fixed base simulator (FBS) simulated the on-orbit operations of the Space Shuttle and included the flight deck, the middeck, and even the potty! On the flight deck, high-fidelity switch panels, computers, and visual simulations of the mechanical arm, rendezvous, on-orbit navigation, docking, maneuvers, and any operations required to be done by the crew while on orbit could be simulated. On the middeck, the locker configuration and locker contents, galley (kitchen), switch panel, and other equipment needed for a mission were simulated. During a simulation in the FBS, the flight procedures in the FDF were used, and the flight configuration was as close as possible to the actual mission. Our trusty instructors had a large control room where they could introduce a malfunction with the stroke of a light pen on their consoles!

The motion base simulator (MBS) simulated the launch and entry operations of the Space Shuttle and only included the flight deck of the orbiter, with seats for the commander (left seat); pilot (right seat); mission specialist two, or MS2 (middle seat); and another mission specialist (behind the pilot and beside MS2). Yes, the MBS simulated the launch! With the switches and panels of the flight deck and simulated visual displays out of the windows, the MBS also provided the physical and aural cues of an actual launch or landing. The astronaut would be strapped in while sitting upright, then would be rotated to the launch position where they would be resting on their back. Then, the games would begin, and the Space Shuttle would shake, rattle, and roll! Once again, our Space Shuttle instructors would take us through the wringer with the malfunctions they introduced.

The purpose of the FBS and MBS was to learn how the Space Shuttle systems interacted with each other in the event of a failure and how the procedures should be followed to stay out of trouble. It took quite a bit of understanding of the systems to know what to do and what procedure to follow. It was also important to learn how to work as a crew and share responsibilities. I spent hundreds of hours in these simulators and was always humbled by them at the end of a simulation!

My father had similar experiences in the Link trainer used to simulate his B-17 or other aircraft. It was an MBS as well, but it was not as complicated as the Space Shuttle MBS. I never heard him talk about it, but I am sure that he was humbled by this "Blue Box," as they called it, because it was used to simulate flying in instrument weather conditions (*i.e.*, in the clouds).

Our class continued to have many social outings together over these months and really enjoyed the camaraderie. We were all settling into Houston and into our new homes, so it was fun to entertain. We mission specialists also became more and more familiar with the pilots in our class and enjoyed outings in the T-38s with them. Somehow, we found time to take a ski trip together in February 1988 in Taos, New Mexico.

Training continued with simulators and more simulators and with T-38 flying whenever I had the chance. It was the best job I could imagine, and I was just getting started!

The Astronaut Office was abuzz because everyone was excited about the prospect of flying the Space Shuttle again after the Challenger accident. The return-to-flight crew for STS-26 was very busy getting ready, and we were all watching the remedies that were in place that would hopefully result in a safe mission. The next flight crew for a Space Shuttle mission was called the "prime crew," and they received priority for training, flying T-38s and managing other aspects of getting ready for a spaceflight. We were all delighted to watch the prime crew training to prepare for their mission.

A new chapter in my life as an astronaut was about to start. After months of training, including taking a workshop on public speaking and media training, we were authorized to go out into public and make speeches. I was never bothered by speaking to groups, so I was looking forward to the opportunity. I didn't understand who would want to hear me talk, as I had not even flown in space yet! However, there were not many female astronauts, so we were in high demand. We were expected to give one public speech a month, and I gladly complied with this requirement. I had the opportunity to go all over the country and talk about space. It was a good gig!

I decided that I needed to do other things with what little free time I had, so I started volunteering as an assistant troop leader with the Girl Scouts. I worked with a troop that my classmate Jim Voss's daughter was in, and it was the perfect outlet for me. As a Girl Scout, I'd always admired the organization, the Girl Scout Law, and the things that they emphasize: God and country. So, I helped the Brownies (who later became Juniors) with science badges as well as camping out and field trips.

I also did things like taking flower arranging classes, exploring different parts of Texas, and traveling. I had relatives all over the state, and I took the opportunity to see them when I could. I also visited Daddy Ben, who was still living in Fort Worth. When I visited him there, I again saw Holly who was all grown up with a toddler, Krystle. It was during this visit that I first saw the many stagecoaches and wagons Daddy Ben had made. He was very proud of them, and for a good reason!

While I was there, Daddy Ben showed me his Wartime Log and talked about his experiences as a POW for the first time. That was the only occasion he ever talked to me about it, and he explained his ordeal as we slowly turned the pages of his Wartime Log. I was struck by the beauty of the paintings and drawings and was impressed again by his many talents.

In August 1988, our class was responsible for the astronaut reunion, at which we would receive our silver astronaut pins. I was in charge of decorations, and there were balloons, balloons, and more balloons! It was amazing to meet some of the former astronauts, including Wally Schirra and some of the Apollo astronauts. What an experience! It was also an honor to receive my silver astronaut pin. It meant that I was no longer an astronaut candidate but a full-fledged astronaut! It was akin to my father receiving his silver wings. Although we had worked hard for those silver decorations, more training and flights were required before we were ready to go on our missions.

September 1988 was a joyous occasion in the office, as the Space Shuttle returned safely to flight on STS-26. It was a great morale booster for NASA and especially for the astronauts. We were flying again, and we were all full of hope and excitement for future spaceflights (and future spaceflight crew assignments!) I was also relieved to know that my ETA redesign on the SRB worked well and would be used on the Space Shuttle for the rest of her flights.

For the Astronaut Office's 1988 Christmas party, our class was once more responsible for the entertainment. I remember that entertainment very well, as we performed several skits, and I was the Santa Claus! I disguised myself effectively with gloves, sunglasses, and a Santa suit (Figure 17-9). Some people approached me afterward and asked who the Santa Claus had been!

Different Astronaut Office managers were asked to come up and sit on Santa's lap, and I had questions for them prepared in advance. The head of the Astronaut Office at the time was Dan Brandenstein, and one of his duties was to help with the assignment of the astronauts to upcoming spaceflight crews. When he came to sit on my lap, I ad libbed the question,

"You don't know who I am, do you?" Many told me that was the best line of the whole party! It brought the house down.

Figure 17-9: Jan disguised as Santa Claus for the Astronaut Office Christmas party, pictured with astronaut Steve Hawley (1988)

Most of the astronaut classes had nicknames, and eventually, our class unofficially became the "GAFFERS." GAFFER stood for "George Abbey's Final Fifteen." As you may recall, George Abbey was head of the Flight Crew Operations Directorate and was a very influential member of the astronaut selection board. He was assigned to headquarters in March 1988, so we thought we were the last astronaut class he selected. However, we were wrong. He may not have been on another astronaut selection board, but he stayed in the space business as a prominent leader and became center director of JSC in 1996. He was always very involved with the astronauts and human spaceflight, so we did not make our name very widely known!

We were fortunate that the Space Shuttle continued to fly safely, and morale and spirits were high in the Astronaut Office in 1989. People were being assigned to spaceflights, and there were many celebrations—splashdown parties, promotion parties, and crew assignment parties at the Outpost Tavern. Our astronaut class was being divided up to work various jobs within the Astronaut Office, and I missed the frequent get-togethers we'd had before we received different assignments.

Early in October 1989, I was surprised to be called into the office of the chief astronaut, who at the time was Mike Coats. He asked if I would be interested in going to Japan and being a part of the first Space Shuttle crew STS-47 with a Japanese. This would mean training in Japan for about two and a half months. I could not say, "Yes, of course!" fast enough! A bonus was that my classmate, Mae Jemison, was also assigned to this crew. The flight included a Spacelab module, and the Spacelab training would be conducted in Huntsville at MSFC.

It was an exciting day for our astronaut class, as five of us were assigned to spaceflights—the first group from our class to be assigned. We had the requisite celebration party at the Outpost, and I made a lot of phone calls that evening. I didn't know when or where I would be going, but I knew that I was on my way!

Chapter Eighteen
Fuwatto

Crews for a spaceflight typically were assigned about one year prior to the scheduled flight. Because of the number of experiments and the amount of travel for a Spacelab mission, the mission specialists were assigned two years or more in advance. When we were assigned to Spacelab J (J for Japan) on that day in October, our flight was scheduled to fly in mid-1991.

The payload crew for Spacelab J was to consist of two mission specialists and two payload specialists. Payload specialists are astronauts who are selected to be on one specific flight for the experiments or payload on the mission. One of the payload specialists would be chosen by a cadre of scientists who had the American experiments, and one payload specialist would be chosen by the Japanese.

The Japanese elected to select three payload specialist candidates, Takao Doi, Mamoru Mohri, and Chiaki Mukai. A down selection to one payload specialist who would fly on Spacelab J would be made by the Japanese closer to the flight date. For the American payload specialist, NASA chose to select a mission specialist who met the criteria of the payload specialist and who would be approved by the American group of scientists who had experiments on the flight. Therefore, a new term was coined, "science mission specialist," and this person took the place of the slot previously held by the American payload specialist. Mae Jemison, a medical doctor with an undergraduate degree in chemical engineering, was selected for this role, and the previously selected American payload specialist, Stan Koszelak, would be her backup.

Mark Lee and I were the two mission specialists assigned to the flight who would complete the payload crew. Mark was in the class of 1984 and had flown once before. He was named the "payload commander," a new term that denoted the person who oversaw the payload portion of the mission, including training, FDF, assignment of duties, scheduling, and whatever it took to make the mission successful.

The Spacelab missions were managed by MSFC, and the Spacelab module training (not including the Spacelab J experiments) would be done at the Spacelab simulator in the Payload Crew Training Center in Building 4612. Our training manager from MSFC was Homer Hickam, whom I knew from diving in the NBS at MSFC. Homer would accompany us to Japan and for all our Spacelab related training.

As we traveled to Japan in October 1989, one of the highlights of our training was meeting and working with the three Japanese payload specialist candidates and the backup American payload specialist. A photo of some of the payload crew at a press conference is shown in Figure 18-1.

Figure 18-1: Payload crew for Spacelab J, STS-47.
Front row: Jan, Mamoru Mohri, Mark Lee.
Back row: Takao Doi, Mae Jemison, Chiaki Mukai.
Missing from photo: Stan Koszelak

For most of our initial science briefings, we were based in Tokyo. I had my ice skates, so I was able to skate at all the Tokyo ice rinks. Mae and I really enjoyed experiencing Japanese culture and went to museums as well as kabuki and bunraku theater.

A big adjustment for me was the food. I'd had Japanese food before but not the variety that I had in Japan. Once I became used to it, I really enjoyed it, especially sushi and sashimi, curry rice, teriyaki, tempura, katsuju, noodles soup (udon or soba), yakitori, delicious pastries, okonomiyaki, takoyaki, sukiyaki, shabu-shabu, and Pocky—chocolate-covered pretzel sticks. You may be familiar with many of these dishes that are served elsewhere in the world.

Almost every night, there was an event with lots of toasts and eating. We met many Japanese dignitaries, and hopefully we were polite in our greetings, in our bows, and while drinking our green tea. The Japanese called our mission "Fuwatto," which means "floating" or "weightlessness" in Japanese.

The twenty-three-foot-long Spacelab pressurized module was a joint project with NASA and the ESA. It was a wonderful, highly complex facility that had its own environmental control for the humans and experiments, data and television systems, communication systems, and heating and cooling for the experiments. About the size of a large school bus, it had the capability to house experiments in the racks and drawers for storage in the ceiling. For Spacelab J, the configuration of the module had eight double racks and four single racks of equipment. These racks contained the furnaces, workstations, experiment facilities, storage compartments, and support equipment needed for the majority of the experiments. The module was attached to the middeck of the Space Shuttle orbiter by a sixteen-foot-long tunnel, where the astronauts could "go to work" every day in a comfortable shirt sleeve environment.

On Spacelab J, there were forty-three experiments, of which thirty-four were sponsored by Japan, in the areas of materials science and LS. With my degree in biology and my graduate research in the field of materials science, I was comfortable with the science on the mission and developed a trust and mutual respect with the principal investigators.

One might wonder why we go to space to perform these experiments. The environment is called "microgravity" because, technically speaking, there is still gravity experienced from the Earth. However, the Space Shuttle is in a state of "free fall" as it orbits the Earth, so if you put a g-meter inside the Space Shuttle, it would read zero. It's as if you were on an elevator, and the cable broke. You would still be experiencing the effect of gravity, but you would be floating in a state of free fall (at least until you hit the ground!) So, we can do basic science in space and not have the apparent effects of gravity affecting the experiment because it is experiencing zero-g inside the Space Shuttle.

After the initial science briefings, it was time to start training on the hardware, which was contained in the experiment rack simulators. Our first visit was to the facilities of the Ishikawajima-Harima Heavy Industries (IHI) just outside of Tokyo in Tachikawa. IHI was responsible for building the two double experiment racks for the Japanese Materials Experiment Laboratory (MEL), and their high-fidelity simulator was located at their facility. The Japanese instructors did an incredible job of writing down very detailed procedures and drawings about how the materials science experiments would be conducted.

Figure 18-2: Training on life sciences experiment rack in Kobe. Pictured: Mark Lee, Stan Koszelak, Jan, Chiaki Mukai, Mae Jemison, and training manager Homer Hickam

After a few more weeks in Tokyo, it was time to go to a different location to train on the LS experiments. The Mitsubishi Heavy Industries (MHI) was responsible for building the double rack that housed the Japanese LS experiments. Figure 18-2 shows the LS rack at MHI along with some of the payload crew. The MHI facility was a large shipbuilding plant in Kobe, and we took a bullet train to get there. That was a thrill in itself! Although a large, industrial city, Kobe had many beautiful mountains and countryside, which was a welcome change after being in a big city for several weeks.

Mae and I spent a weekend together going to the beautiful city of Kyoto where there are very old temples that were spared from the bombing during WWII. We visited and appreciated the culture, the art, and the famous landmarks in Kyoto. One of the principal investigators for a glass experiment on Spacelab J was at a university in Osaka, so we were able to see some of the landmarks there as well.

The training at MHI in Kobe was moving at a good pace, and the LS experiments were very interesting. They included live subjects, such as an experiment with koi (goldfish). It was not possible to train for everything on the ground, as some things would act differently in space, but we did as much training as we could, and the simulators and instructors were fantastic.

After a few weeks training in Kobe, we returned to Tokyo and went to Tsukuba, a modern city that houses all the scientific and technical agencies in Japan. It was very impressive, and I thought it was a good decision to have major test facilities, science facilities, and technology development in one location. The Japanese space agency (NASDA) center was in Tsukuba and included good facilities for training crew, including a water tank for EVA training. The human spaceflight program in Japan was well on its way, and they were ready to recruit more astronauts to do their preliminary training in Tsukuba.

We were to go home on December 23, and I was ready! I had done some Christmas shopping, and everyone was either getting Japanese cookware, china, or jewelry for gifts.

When I returned to JSC at the beginning of 1990, our training on the Space Shuttle and much-needed flying in the T-38 continued.

Our next trip to Japan was in March for one month, and the trip was scheduled beautifully by the Japanese training team. It was during the highlight of the cherry blossom festival! I had seen cherry blossoms in my hometown of Huntsville and in Washington, D.C., but the experience in Japan was more than just the observance of the flowers. We were in Kobe training at MHI during the height of the blooming of the blossoms, and everyone had picnics after work under the cherry trees. It was quite a festival, complete with food, beer, and karaoke machines! The cherry blossoms were beautiful and looked like snow when they blew off the trees. It was a fun outing with friends and family.

When you are in Japan, you must visit Mount Fuji! It was a long trek by train from Tokyo, but one weekend day, we went to a resort in Hakone with our whole American gang. It was a wonderful day with great weather, and the views of Mount Fuji were spectacular. It is easy to see why it is a national symbol for Japan.

After returning to the U.S., it was more T-38 flying and Space Shuttle proficiency training in the simulators, and we started traveling around the country for our scientific experiments on Spacelab J that had American principal investigators. Most of our time was at Ames Research Center, a NASA center that is known for its LS expertise. We also had training at JSC for their LS experiments and at Glenn Research Center for a technology experiment.

Later in 1990, it was decided that the United States Microgravity Laboratory would jump in front of us and that we would be delayed another year. When I received this news from the head of the Astronaut Office, he told me that I could have any job in the office for the year that I was not training. I told him I wanted to be a capsule communicator (CAPCOM), a member of the mission control team and the only person who communicates with the spaceflight crew in orbit. As a result, I was the CAPCOM for six Space Shuttle missions—five in 1991 and one in early 1992.

Also during my year off from Spacelab J, I was tasked to become intimately familiar with the Japanese materials science experiments in the MEL racks. Along with that, I kept up my proficiency in the Shuttle Mission Simulators (SMS).

I continued to fly T-38s and was also working on my instrument rating as a private pilot. In addition to all of this, I did many speaking engagements. The space program belongs to the American people, and it is nice to go out and share their space program with them. The public appearances were usually upbeat and positive, and I still enjoy doing them.

Daddy Ben was very interested in my training and my flying. We corresponded frequently, although I wish now that I had traveled to Fort Worth from Houston to visit him more often. He was busy taking care of his older siblings and occasionally his granddaughter but found time to continue his hobbies. He built more stagecoach and wagon type models and also built large model aircraft from scratch instead of building them from kits (Figure 18-3).

Figure 18-3: Ben Smotherman pursuing another passion—building model airplanes from scratch

As my CAPCOM duties came to a close in early 1992, it was time to start training again for Spacelab J. The rest of our crew was named, and we were assigned Hoot Gibson as our commander, Curt Brown (one of my astronaut classmates) as our pilot, and Jay Apt as our flight engineer. The rest of the crew had trained extensively in Japan and in Huntsville on the Spacelab J payload , but as a complete crew, we started training to integrate the Spacelab and Space Shuttle activities.

Along with the announcement of the remainder of the United States portion of the flight crew, the Japanese selected the Japanese payload specialist who would fly with us: Mamoru Mohri. Chiaki Mukai and Takao

Doi would be his backups if he could not fly. All three of them moved to the United States for about a year so they could be there full time for Space Shuttle training.

In Houston, we practiced emergency scenarios in a variety of ways. In our orange LESs, affectionately known as "pumpkin suits," we were going to be strapped into our parachute and our seats for launch and entry. In addition to our parachute, our LES had a "survival backpack," which contained a life raft, survival gear, a thirty-minute supply of breathing oxygen, and a water tank.

In the WETF, we learned how to use some of our emergency equipment in the water. The big and bulky pumpkin suits made maneuverability difficult, and they were very hot. Nevertheless, we learned how to get out from under a parachute and how to deploy and get into our life rafts.

Jay Apt and I were assigned as the EVA crewmembers. Jay had been on EVAs before, and it was exhilarating for me to do the training with him underwater in the WETF. There were no EVAs planned for our mission, but we practiced and trained for contingency EVAs. For example, if the payload doors would not close while trying to prepare the orbiter for reentry, there was a contingency EVA procedure for operating the latches that would enable the closure of the doors.

An important activity for our newly assembled crew was taking a professional photograph. What most people don't know is that there is also usually an unofficial crew photo. We decided to take ours in Japan, where the Americans would wear Japanese costumes, and the Japanese would wear American costumes. It was great fun (Figure 18-4), but I must admit that the kimono was very uncomfortable! After wearing one, I had even more admiration for the Japanese women who wear the kimonos with grace and beauty.

So, our seven-person crew was complete, and we finalized our Space Shuttle training in the simulators at JSC and our Spacelab J training in the simulators at MSFC. We also did joint integrated simulations that included the mission control team at JSC and the payload operations team at MSFC. During our Spacelab simulations in Huntsville, we interacted with the Spacelab Mission Operations Control Center at MSFC, as they had a full control room to monitor and control the activities back in the Spacelab and were fairly autonomous from the mission control activities.

Figure 18-4: Spacelab J, STS-47, informal crew picture (NASDA photo)

As we drew closer to our launch date of September 1992, there were the ordinary things for a crew like autographing crew pictures, making up guest lists, and selecting a few items for our official flight kit (OFK) and personal flight kit (PFK) that would be flown in space. In addition to mundane things like choosing our personal items (toothbrush, toothpaste, deodorant, etc.), we also ordered our personal shirts and other clothing. One of my favorite things to do before a flight was to select our food after a food tasting test.

After giving Daddy Ben several choices of what I would fly into space for him, he selected a beautiful gold Space Shuttle tie tack. Of course, he and others in the Smotherman family, the Davis family, and my mother's (Gantz) family would be invited to the launch.

The summer of 1992, I received a message from Darby telling me that Daddy Ben had lung cancer. Evidently, the many years of pipe and cigarette smoking had caught up with him. So, a T-38 pilot and I flew to

San Antonio, where Daddy Ben was at the Lackland AFB Wilford Hall Medical Facility.

When I visited him there, he was upbeat and looked good, and Darby was in the room with him. It was wonderful to see them, and he and my T-38 pilot were interested in exchanging flying stories. I don't think many Air Force pilots had the opportunity to talk with a WWII pilot, much less one who had been a POW. They both seemed to really enjoy the conversation.

After Daddy Ben and I caught up on the latest news with each other, he wanted to get up and walk around. So, with his IV still attached on a wheeled tower, we walked the halls of the hospital. It was a heartwarming visit, and I am so glad that he was able to see me in my astronaut flight suit. Plus, I am thankful that the Air Force hospitals in San Antonio took such good care of Daddy Ben when he had his lung cancer.

I was participating in many SMS simulations, in addition to my other training, especially entry simulations. During launch, it was planned for me to be in the middeck, but during entry, I was supposed to be on the flight deck.

During one of the simulations, the control room for the SMS asked me to come out of the MBS into the instructor's control room. The lead instructor told me they had just received a phone call saying that Daddy Ben had died. It was July 2, 1992. I was stunned but told them that I would continue the training in the simulator, as that is what Daddy Ben would want me to do.

Due to my schedule, I was not able to go to the funeral—another regret. I was able to communicate my condolences to the family, and they understood why I could not come. Daddy Ben was seventy-three when he died. There are so many questions I wish I had asked him.

As the launch date in September grew nearer, the training pace quickened. Much more training was added in the last few months so that we were sure to remember everything. While on the Space Shuttle, the flight crew had many cameras at their disposal for interior photography as well as external photography of the Earth. After receiving training on how to use them, we were able to check out the cameras and camcorders to become familiar with them.

Between launch familiarization at the Brooks centrifuge and flying T-38s to Kelly AFB and Randolph AFB, I became familiar with the southwestern styled buildings and the airfields at the San Antonio Air Force bases. I did not know at the time that my father had received his initial training as an AAF cadet there. It makes it more special that I, too, trained in San Antonio, although in a different way. The last place I saw him was in San Antonio at the hospital where he died. So, he came full circle—his first training was there as an aviation cadet and his last breaths were taken there.

In preparing for the launch of the Space Shuttle, we had a dress rehearsal of the countdown with our spacecraft Endeavour on the launch pad at KSC. As we suited up in our orange LESs in the suit room, I was in awe of the fact that this same room had been used to suit up every crewmember that had ever flown in space for NASA. We sat in reclining easy chairs while the technicians inflated our suits and checked to make sure they could hold pressure. Our communications cap (which we called the Snoopy cap because the astronaut Snoopy wore one!) was checked out, along with our helmets and gloves.

As we drove out to the launch pad in the Airstream trailer, called the "Astrovan," I thought about how many astronauts who had flown into space had ridden this trailer to the launch pad. Meanwhile, we were chatting without the nerves that would be there on the actual launch day. When we arrived, we exited the Astrovan and went up the clunky elevator, which was obviously the original. After getting off, we could walk around the launch pad, and at that high level, the view of the vehicle and KSC was magnificent. We walked down the gantry and into the creatively named White Room where the technicians strapped on our parachutes and made sure everything looked okay. Then, we ungracefully crawled through the side hatch into the Space Shuttle.

The Space Shuttle is vertical on the launch pad, meaning our seats were ninety degrees from where a normal seat would be. Therefore, we had to use all kinds of straps and gadgets to pull ourselves into the seat with the help of technicians and the astronaut support personnel (ASP).

Finally, I was lying on my back in my seat on the middeck. Whew! That was hard work considering we were in the Florida summer heat, and I had no cooling. It was a welcome relief for the suit technicians to hook up the fans from the Space Shuttle into the fitting on my suit, although they did not provide much cooling because they were circulating ambient air.

It was somewhat disorienting to be on our backs; for some reason, the switch layouts on the panels and the hardware seemed different. It was strange to be rotated ninety degrees from the orientation that we had in the simulators.

Again, it was as if I were having an out-of-body experience to realize that I was strapped into the actual Space Shuttle that I was going to ride into space. What a thrill! In addition to mentally reviewing the procedures we had to follow, I looked around at everything in great detail and pinched myself that I was really there.

Once everyone was strapped into their seats, the dress rehearsal, called the terminal countdown demonstration test (TCDT), began. Our communications were with the launch control team at KSC, who were at their consoles in the Launch Control Center (LCC). On board Endeavour, the procedures and activities that would occur the morning of launch were coordinated by the LCC and performed by the flight crew. The total time of the TCDT was about two hours, which is the amount of time a Space Shuttle crewmember would typically be on the launch pad before liftoff. As I was on the middeck, along with Mae and Mamoru, I had very little to do since most of the crew activities for TCDT were performed on the flight deck. Three… Two… One… And no liftoff—thank goodness. We then followed our procedures and flipped the necessary switches to make the Space Shuttle safe in a simulated emergency.

After unstrapping and unhooking our suits from the cooling fan, an oxygen supply, our parachutes, and our communication line, we left our seats and opened the hatch that was in the middeck. Then, we ran out into the gantry and simulated hitting paddles that activated a water deluge system. With the nozzles spraying water, it was impossible for the escaping crew to see anything but their feet, so on the open grating, there was a bright yellow path, which pointed to the slide wire baskets. We fondly called this "the yellow brick road."

If there was a fire or pad abort on launch day, the flight crew would jump into baskets (two per basket) and ride them down to the ground on a slide wire. After climbing into the baskets, we would hit a paddle that would send us plummeting. For this dress rehearsal, the baskets were chained to the launch pad, so we weren't supposed to go anywhere! As a result, we didn't practice going down the slide wire, but we did practice jumping into the baskets in our bulky LESs. After riding down to the ground, we were to hop out of our baskets, run to the green M-113 armored personnel carriers, and drive them away as fast as we could!

During this extensive training for contingency scenarios, I often wondered what type of training the WWII aircrews had received. They'd had a limited amount of parachute training, and every crewmember had known how to exit a failing B-17 (Figure 18-5). My father, as the pilot, would have rung a bell and ordered "bailout" before jumping out of the bomb bay door. As in a Space Shuttle emergency, a bailout command in a B-17 was not always an orderly process due to high g's or an uncontrollable aircraft.

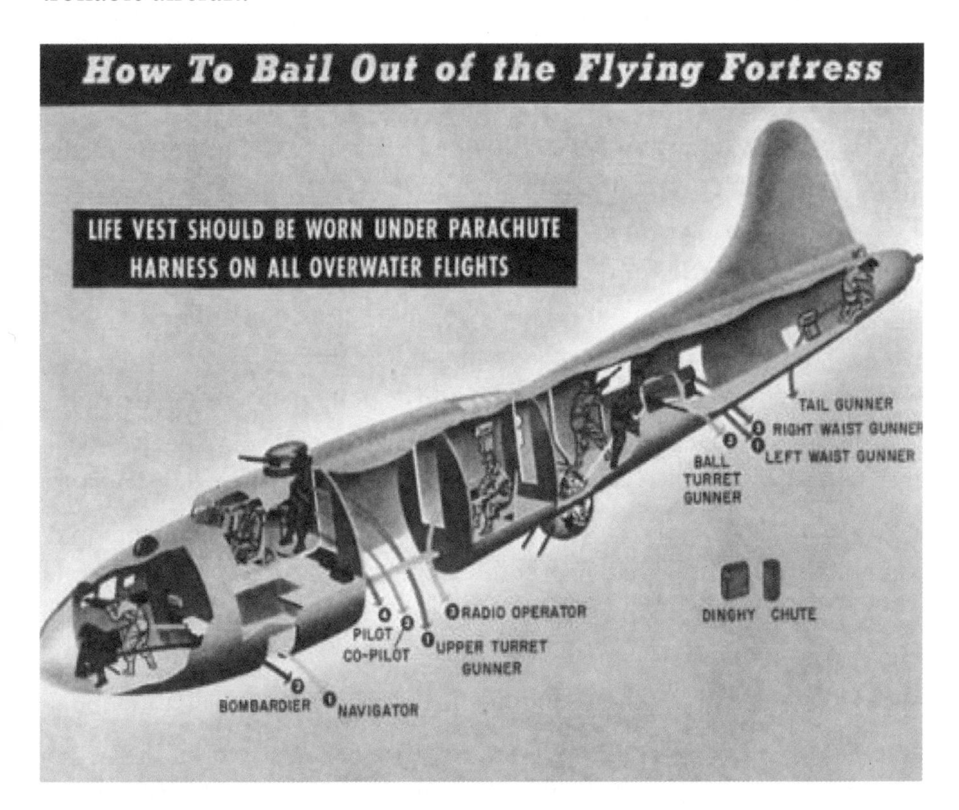

Figure 18-5: Bailout paths for B-17 crew (USAAF training diagram)

The B-17s were pretty good at sustaining being ditched, as proven by one of my father's colleagues, Bill Stewart, who had to ditch his B-17 (Chapter Four). The aircraft had been equipped with dinghy life rafts, so they'd received training on how to deploy them, just as we'd learned to deploy and inflate our dinghies for both the T-38 and the Space Shuttle. Our training was very thorough, but preparations for flying in WWII had been fast and furious, and I doubt that they'd received much emergency training at all.

Everything was looking good for a launch in September, and our families and friends were sending their RSVPs for access to view the launch at KSC. A week before a Space Shuttle mission, the spaceflight crew was put into quarantine at the crew quarters at JSC. Our meals were prepared by JSC staff so that no illnesses from spoiled food or whatever could occur. In addition, the flight crew could have no contact with children; we could only have contact with adults who had been examined and cleared by a NASA flight surgeon. The crew quarters had all the cameras that we used in training, all flight procedures, and any other documentation that we might want to study.

Three days before we were to launch, we flew T-38s as the "prime crew" (the next crew slated to launch). We flew in formation with four T-38s as our chariots going to Florida. Just before we landed at the Shuttle Landing Facility (SLF), we did a fly-by of the launch pad. It was riveting to look at—that was our rocket! We were still in quarantine and were isolated once again in the crew quarters at KSC.

Two days before the launch, we could each invite five people to have a barbeque with us at the KSC Beach House, an old, small beach house that had been refurbished (Figure 18-6). The expanses of sea and sky made it the perfect place for those of us who were about to launch to say our goodbyes to our loved ones before being separated by space and time.

My mother never talked about it then, but I know that she was scared and happy at the same time for me to be going into space. I'm sure that she'd experienced a similar feeling when she'd said her last goodbye to my father before he went off to war. In both cases, that is what we wanted to do and what we were trained to do. It was not only our vocation, but it was our passion and a service to our country. For those reasons, my mother fully supported both of us in our endeavors. Nevertheless, it was hard to say goodbye to her at the barbeque at the beach house. No one said it, but everyone knew that it could be the last time that we saw each other.

Figure 18-6: Refurbished beach house at Kennedy Space Center after being made into a conference center (NASA photo)

Amid studying cameras, reviewing procedures, and staying safe in crew quarters, we could not help but think about the upcoming mission. I really did not think about the risk or the danger of launching into space—I just wanted to do a good job and not mess up any of the experiments that the scientists had worked so hard to develop. I believe that these were the same thoughts that my father and others had going into combat. Although the danger and the loss of their friends was in the background of their minds, in the forefront was how to execute the mission as best they could.

After having seen my family at the beach house two nights before my launch, the time before launch was a bit melancholy. I was deep in thought about what I was about to do and for what I had trained for almost three years. The countries of Japan and the United States were counting on us to do well, and I did not want to let them down. Isn't this what every pilot in a war wants to do—perform for their country? It was a privilege, not a burden, but it was on my mind very much, just as it was for those airmen going into war. That is what we all signed up to do, and that is what we were willing to accept.

The day before launch, I went for a jog to get some fresh air. I also tried to eat healthily and not break my leg or something like that. Overall, I decided to relax because I thought if I did not know something by now, it was too late!

It was customary to have a prelaunch party one day prior to the launch, and I had planned this party at the PAFB Officer's Club, which was a nice facility on the beach. I thought it was neat that I had come full circle—from being born at PAFB to launching from a nearby facility that had once been the Canaveral AFS where my father had tested Matador missiles. It was the same Officer's Club that my mother and father had used in the early 1950s, so it was very nostalgic for me to have my prelaunch party there. Unfortunately, I was not allowed to go to the party and enjoy seeing everyone, as I was in quarantine!

Due to our sleep shifting, the blue team (of which I was a part) went to bed at 10 a.m. and woke up at 6 p.m. That was hard to do! When I awoke, the prelaunch reception I was hosting at PAFB had started, so I gave them a quick call on the speaker phone to thank them all for coming to the launch. The red team went to bed at 9:30 p.m.—a respectable time!

The red team was awakened early in the morning on September 12, 1992, and we all got ready for a 10:23 a.m. launch. I had butterflies in my stomach—a little bit from nervousness but mostly from excitement.

You have probably seen pictures of Space Shuttle crews around a table at breakfast time with a big cake that has the crew patch in icing on the top. Well, that is all for show! We were all smiling behind the cake for the cameras, but we didn't eat it for breakfast. After the cameras left, someone took the cake and put it in the freezer for us to enjoy after the mission.

From there, each person picked out his or her own breakfast. Some astronauts had the traditional "steak and eggs"; some had a big southern breakfast of bacon and eggs; and others had a small meal of cereal, avocado toast, or something else. As I did not know how my body would react to space, I decided that I would keep it simple. My normal morning repast was a bagel with cream cheese and coffee, but I just had a chocolate protein drink replete with vitamins and nourishment.

After that, it was time to get suited up, and we went through the same process that we'd used during the dress rehearsal to get ready. The suit-up room was such a historic room, and again, I could not believe that I was (hopefully) going into space that day! After suit-up, nerves were relaxed by playing cards just before we walked out of the Operations and Check-out Building, the home of KSC crew quarters. This day was different—there was a crowd of people cheering us on as we walked out to the Astrovan at 7 a.m. (Figure 18-7). Some were familiar faces, some were our instructor team, and all played a part in making sure that all systems were go for a launch. It was a beautiful day in September, and life was good.

Figure 18-7: Spacelab J, STS-47, crew walks out of crew quarters to the Astrovan waiting to take them to the Space Shuttle launch pad, September 12, 1992 (NASA photo)

As we were riding to the launch pad, the chatter was lighthearted and funny, as we were all trying to relax. Our commander, Hoot Gibson, said, "The only thing that I want to say is... Don't screw anything up!" Again, riding the same Astrovan that had been used since the early days of our space program was humbling to me. I finally realized that this day was really happening—it was not a dream.

When we approached the launch pad, there was no one around, unlike other days when we'd been at KSC. The Space Shuttle had been loaded with fuel and ordnance, so it was a dangerous vehicle that only a few people could get near, including those of us who would fly on it. As we disembarked the Astrovan, Endeavour was alive. It was creaking and groaning, venting off gases in clouds of white and letting us know that she was ready to fly.

As we once again rode the elevators to the 195-foot level, I thought about how many flight crews had ridden that elevator on their way to their respective launches. It was old and rickety and slow, and I wondered if we would make it. These emotions were very different than any that I had ever experienced before while riding up the elevator in the launch pad. I had a lump in my throat just thinking about those who had gone before me.

When we reached our destination, we looked around a bit but then walked across the gantry to the White Room, carrying our helmets. Today was not a day for sightseeing, and our timeline was very tight. It was time to fly into space, and no one was more eager than I.

Chapter Nineteen
Off I Go...

The White Room was the area just outside of the Space Shuttle hatch where suit technicians and Space Shuttle hardware engineers prepared the crew and their suits for launch. They helped us put our heavy parachutes on and made sure that our caps for communication were snug and secure.

I climbed into the hatch on my hands and knees and waved my last goodbye to the camera. The guys on the flight deck had already been loaded, and then it was my turn. Mae and Mamoru were loaded into the middeck after me. As soon as I had my communication headset connected, I said a hearty hello to the rest of the flight crew on board to do a comm (communication) check. They were busy going through their procedures, and I tried to keep the intercom loop free of chatter. The suit technicians and the ASP gave my helmet, gloves, kneeboard, etc., to me as they strapped me in. The ASP for our flight was Eileen Collins, one of the first woman Space Shuttle pilots.

I remembered climbing into the B-17, which was also cramped and designed for skinny young men in their twenties. They, too, had bulky jackets, parachutes, and equipment that they hauled on board. Their strapping-in process may have been a little less laborious than mine, but the nerves were undoubtedly the same.

We were all nestled into our seats when the support personnel took out the wooden platforms they were standing on and closed the hatch. Then, they were gone. We were now all alone out there on the launch pad—just us and Endeavour. Those of us on the middeck did not have any duties during the two-hour launch countdown, so we were left alone with our thoughts while the flight deck crew readied us for launch.

Meanwhile, our family bus was offloading at bleachers at the Banana Creek viewing site, and our other friends who had causeway passes were making the trek to park their cars on the causeway at KSC. Immediate families (spouses and children) were in the office of Norm Carlson, a Space Shuttle manager, located in the LCC.

As the countdown clock neared nine minutes before launch, or T-9 (time minus nine), there was a planned built-in hold so the controllers in the LCC could be polled for their final "GO." Also during the nine-minute hold, the immediate family members were escorted to the roof of the LCC so that they would have a great view of the launch. They were very well taken care of by the astronauts who were the family escorts, and medics and security personnel were standing by in case they were needed. No one would talk or think about a bad outcome on that day, but there were strict procedures on what the escorts would do with the family members if there was an accident.

During the nine-minute hold, we all checked our connections and put on our gloves. It was a busy time for everyone as we got ready for the final countdown. We also listened to a quick speech by Vice President Quayle, who came to the launch and congratulated Mae on her historic flight as the first African-American female astronaut and wished us all well.

The countdown clock started ticking down from nine minutes, and it seemed to go very fast. Five minutes before launch, pilot Curt Brown started the auxiliary power units for the hydraulics. I could hear them come up to speed with a whirring noise. It was time for a quiet cockpit—all business now as we were getting down to the critical phases of preparing to go into space.

At T-2 minutes, we were told by the LCC to "close and lock our visors" and "initiate oxygen flow." I pulled down the visor across my face and locked it in place so that it could hold some pressure if our partial pressure suits were needed in the event of an emergency. Then, I flipped the oxygen lever so that oxygen was flowing.

We did one more comm check in a round robin fashion. I was MS3, so I did my check in turn. It seemed like time passed quickly until the ground launch sequencer took over at T-31 seconds. It had been a very smooth countdown with no glitches. The commander exchanged courtesies with the flight controllers and thanked them for all they had done, and the launch director Bob Sieck wished us well.

Those of us on the middeck looked at each other with nervous smiles and gave each other a thumbs up. Before we knew it, we heard a "Go" for main engine start. What? Already? Was I ready for this?

A few seconds before the three SSMEs fired up with their controlled explosion made of a mixture of liquid hydrogen and liquid oxygen, the engines were "gimballed." This is sort of like rotating your arms before a swim meet or making sure that all your joints work before you need them. I could feel the engines gimballing, and my excitement was mounting!

Almost seven seconds before we actually took off, the SSMEs were brought up to full power. As they were cantilevered from the middle of the Space Shuttle stack, the entire vehicle assembly leaned over about a foot. Then, when the vehicle returned to the vertical position (we called it the twang), we were ready to go. I could hear and feel the SSMEs come up to power, and I encouraged them to keep going. Go, baby, go! After all, we'd had situations on two other previous flights where the SSMEs had started and then shut down. That was not only dangerous but also resulted in a short-lived and harrowing launch experience.

Meanwhile, my family and friends could see the bright light of the SSMEs and the steam that was created by the hot exhaust hitting the water deluge system. Just before the SSMEs lit, water shot out below them to suppress the vibration and reduce the acoustic effect. It looked like smoke, but in actuality, it was vaporized water forming clouds of steam (Figure 19-1).

The entire Space Shuttle vehicle was held to the launch pad with eight bolts, four on each SRB. When it was time to launch, a pyrotechnic device sliced the bolts in half, freeing the Shuttle to take off. As I was lying on my back, there was no doubt when the SRBs ignited! Not only did I hear a "POW!" but it felt like someone had given me a big shove—a big, *big* shove—and I was riding down a very rocky road. After that humongous push, it was a bumpy and noisy ride. We were being shaken about a lot—significantly more than our MBS could reproduce. We were on our way right on time at 10:23 a.m.!

Figure 19-1: On-time Endeavour launch of mission STS-47, September 12, 1992 (NASA photo)

Our first communication with mission control was heard shortly thereafter as Hoot relayed, "Endeavour, Houston, Roll Program," and CAPCOM Sid Gutierrez responded, "Roger Roll, Endeavour." During this time, the Space Shuttle pirouetted and oriented itself shortly after clearing the tower to be placed into the proper orbit. I could feel the whole vehicle roll, and this felt unusual, as I could not see anything but lockers in front of me in the middeck.

The medical instrumentation that was monitoring my pulse and heart rate registered a spike at SRB ignition, and then everything returned to normal, and I was cool as a cucumber. Except for the noise and the shaking, the g's and the feel of the launch were very similar to what I had expected.

The shaky, rocky road that I felt I was bouncing around on lasted for about two minutes. Then, there was a bang and a shudder from the SRBs separating from the Shuttle. The crowds on the ground were cheering and whooping and hollering. Some were probably crying, and my mom was likely thanking God for a safe launch. They all happily returned to their vehicles and headed back to their hotels or condominiums for more celebrations. Everyone, especially the flight crew, was ecstatic that the launch had gone off on time! I later learned that this was the first on-time launch without delays since STS-61B in 1985. Wow, we were so lucky!

Back at KSC, the LCC's job during the launch portion of the flight was over when the Space Shuttle cleared the tower on the launch pad. After that, control of the mission was handed over to the flight controllers in mission control at JSC in Houston. Then, it was time to cheer and celebrate at KSC! Honoring a long-held tradition since the first Space Shuttle flight in 1981, the requisite cornbread, hot sauce, and beans were hauled out onto the first floor of the LCC. This was a beloved ritual, and sixty gallons of beans were made every time the NASA team launched another Space Shuttle.

The two SRBs fell to the Atlantic Ocean, and their splashdown was cushioned by parachutes. Then, divers retrieved them, and they were towed back to KSC by boat to be reused on subsequent Shuttle missions.

I could not believe what a smooth ride we had after those SRBs separated from us. The SSMEs felt as smooth as silk and purred along for the next six and a half minutes. As expected, the acceleration increased up to three g's, and I could feel myself being pressed further into the back of my seat. The ET was carrying the liquid oxygen and liquid hydrogen that fueled the SSME, so it was along for the ride as long as the SSMEs were running. When the fuel tanks were empty, the ET would separate from the orbiter, only to break up over the Indian Ocean as it entered the Earth's atmosphere.

"MECO" is what we all wanted to hear—main engine cutoff, when the three main engines would be shut down. After MECO, we could unstrap from our seats, float around, and get down to the business for which we had been training for three years.

At that point, I unstrapped from my seat, and I was floating! It was one of the best sensations I have ever had. It is difficult to describe, but it was fun, exhilarating, and wildly different from anything I had ever done. It was a welcome relief to get out of our orange pumpkin suits! After that, we unbolted our seats and put them away, as we would not be needing them until we came back to Earth.

I was so happy to be on orbit! I felt fine, with no space adaptation sickness (SAS) and no sensation of disorientation or nausea. The first time that Mom saw me on television up in space, she said that I looked right at home. That sums it up, as I was comfortable, happy, and ready to carry out all the wonderful science on our mission.

We had to do some critical things like activate and open the Spacelab module, install the cabin air circulation fan, and complete other housekeeping chores necessary to get ready for the mission. Then, as we all had adjusted our sleep to be split into two shifts, red and blue, my shift—the blue shift—had to go to bed just a few hours after we were on orbit. How hard it would be to go to sleep after so much excitement! To keep us isolated from all the activities of the red shift, we had enclosed "bunk beds" in which to sleep.

Of course, before I went to bed, I had to peek at the glorious Earth from space. We were 190 miles from the surface of the Earth, traveling 17,500 miles per hour, or five miles a second. That meant that we circled the Earth once every ninety minutes, with one half of that time in darkness and one half of that time in daylight. Oh, how magnificent that first glance at Earth was! Neither pictures nor videos do it justice, as the human eye is able to see the Earth with so much detail and so many colors without distortion. It was much more majestic than I could have ever imagined, and I reveled in its beauty. At the same time, I tried to orient myself as to which part of the world I was flying over.

After eating a light meal and changing into some thermal underwear (for pajamas), I floated into my bunk bed. It had a light inside as well as fans to keep the air circulating. I tucked myself into a sleeping bag and strapped myself in so that I would not be floating. Then, I wrote some notes in my journal and recounted the day's events into my small recorder. So much had happened! The next day would be a busy one full of experiments and excitement and exploration.

The second day on orbit, I still felt fine with no nausea or SAS. I did feel congested and had a "puffy" face due to the fluid shift. Without gravity pulling fluids into your arms and legs, the fluid shifts to the body core and head. My experience was typical of spaceflight, and I had been briefed about it. Plus, I'd heard about it from other space travelers. It was supposed to go away after a few days, and it did.

The next day, for the first time, I used the galley in the middeck to eat breakfast. I don't remember what I had to eat that first day on orbit, but one of my favorites was freeze-dried scrambled eggs with some Tabasco sauce. After injecting some hot water into the packet of food, I stuck it in the oven to warm it further. I also prepared hot (well, lukewarm) coffee every morning by using the hot water dispenser.

Those on a WWII combat mission also had scrambled eggs in the early morning for breakfast. If it was to be an all-day mission, the men would get "real eggs," whereas if it was a "milk run" or an "easy mission," they received powdered eggs. I'm not sure whether I had real eggs or powdered eggs, but I thought they were delicious!

After cleaning up, changing clothes, and eating breakfast, I went to the flight deck to look for our daily messages on the teletype machine (yes, I know, very high-tech!) Almost as soon as we'd lifted off the pad, our flight plan had changed, and each day, we had to "cut and paste" a new flight plan into our books. Of course, I could not miss the chance to look out the window at our beautiful world. In fact, if I had any spare minute, I spent it glued to the windows on the flight deck.

Then, it was time to go to work! How many people commute to work by flying down a sixteen-foot tunnel like Superman? There was a little bend at the end of the tunnel before I entered the module, so I could not fly too fast or I would bump my head! I was still clumsily adjusting to zero gravity, but the floating sensation was fun, and it was an environment where it was so easy to work. We could pass large objects, heavy objects, pens, etc., with an ever so gentle little push. It was much easier to handle stowing and unstowing items than it had been in our training in 1-g (gravity experienced on Earth).

Spacelab J was humming and whirring with its fans and machinery. It was alive! It was all too familiar, as we'd had good simulators and high-fidelity equipment for our training in Japan and Huntsville. However, this was not a simulation—this was the real deal!

One of the experiments that I performed on that first day involved melting an aluminum alloy with chopped carbon fiber slivers. In space, we can make materials that we cannot create on Earth because when constituents are melted and floating in microgravity, they can be combined to make a new material. On Earth, the heavier constituent would sink to the bottom due to gravity. This experiment, called "Fabrication of Very-Low-

Density, High-Stiffness Carbon Fiber/Aluminum Hybridized Composites," would produce a very strong and stiff material that we would not be able to make on Earth.

To perform this experiment, I used the continuous heating furnace (CHF), which was an ingenious design. It had four chambers—two heating and two cooling. As it took a lot of power to heat the two chambers to 1,300 degrees Centigrade, two of the chambers were dedicated to heating a sample and were always hot. The other two chambers were used for cooling a sample with the aid of helium gas. When the two samples were finished with their "cooking," they were then rotated to the cooling chambers, and the next two samples were rotated into the heating chambers. In Figure 19-2, I was putting one of the samples into the CHF, and you can also see the bioinstrumentation I was wearing that monitored pulse, respiration rate, and blood flow for a biofeedback experiment.

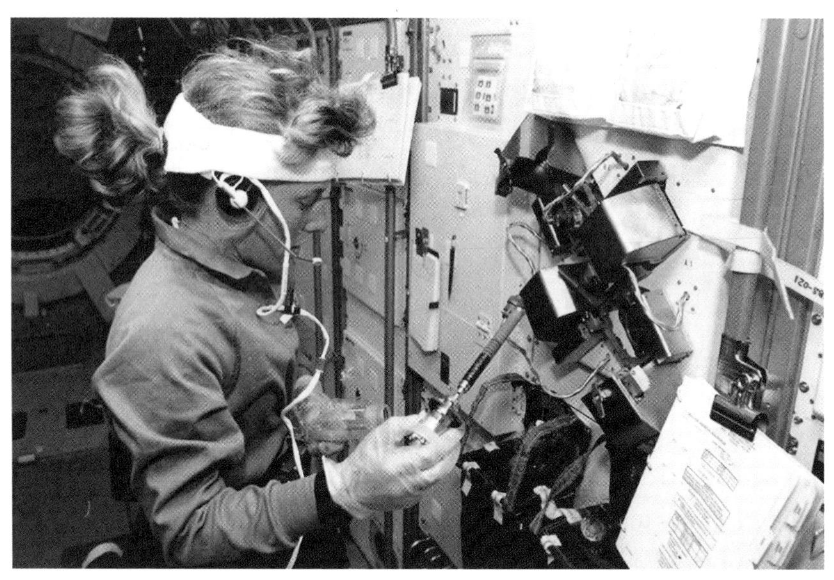

Figure 19-2: Jan operating the continuous heating furnace (CHF) in the Materials Experiment Laboratory (MEL) of the Spacelab J module on STS-47 (NASA photo)

The Spacelab J mission continued smoothly, and eventually, we were able to use the gradient heating furnace and large isothermal furnace after a cooling water leak was repaired. Another of my favorite furnaces was the image mirror furnace. It was a spherically shaped furnace designed to have a floating zone for the sample materials. We'd affectionately called it "splat" during training because in 1-g, we were not able to float any of the samples. It was a beautiful furnace with gold-colored mirrors lining the interior so that the two halogen heat lamps could precisely focus the heat onto the sample.

Four experiments were conducted in this furnace, and one of my favorites was "Growth of Samarskite Crystal under Microgravity Conditions." Due to the high heat required to melt Samarskite and the inability to "grow" a single crystal of it on Earth, space was the only place to obtain a single crystal. From that crystal, the thermal equilibrium drawings required by materials science research could finally be drawn. Good basic science and engineering was being conducted on Spacelab J!

In between my shifts, I spent as much time looking out the window as I could. We were flying at a fifty-seven-degree inclination (so we could fly over Japan), which meant that our orbits covered most of the Earth. We had a bicycle (ergometer) on the flight deck, and this was a good way to get the required one hour a day of exercise and observe the Earth at the same time. My favorite thing to do while watching the world go by was to listen to music. We'd been allowed to take an assortment of cassette tapes, and I found light jazz tunes to be the best accompaniment to watch quietly and thoughtfully what was more spectacular than any video or movie I had ever seen.

The most magnificent things for me to view while floating in space were sunrises and sunsets. Oh my! We were orbiting the Earth sixteen times an earthly day and therefore saw sixteen sunsets and sixteen sunrises (well, except when we were working in the Spacelab!) On orbit, as we passed through our forty-five minutes of daylight, we could see the terminator in the distance—the gradual demarcation of daylight and darkness. As we neared the night sky, the sun started setting on the horizon, displaying a miraculous assortment of colors, highlighting the atmosphere and silhouetting the clouds. In a matter of seconds, the sun disappeared, and nighttime on orbit was upon us. The lights below on the ground gradually appeared as we orbited the Earth. It was really awe-inspiring.

The crew took some phenomenal pictures of the Earth, and one of my favorites was a photo of a sunrise (Figure 19-3). The sun was just visible, peeking over the towering anvil-shaped storm clouds whose silhouetted tops marked the upper boundary of the Earth's atmosphere. Our spaceflight was about a year after the June 1991 eruption of Mount Pinatubo in the Philippines, and the dark gray band that can be seen in the picture is the ash in the atmosphere from the eruption.

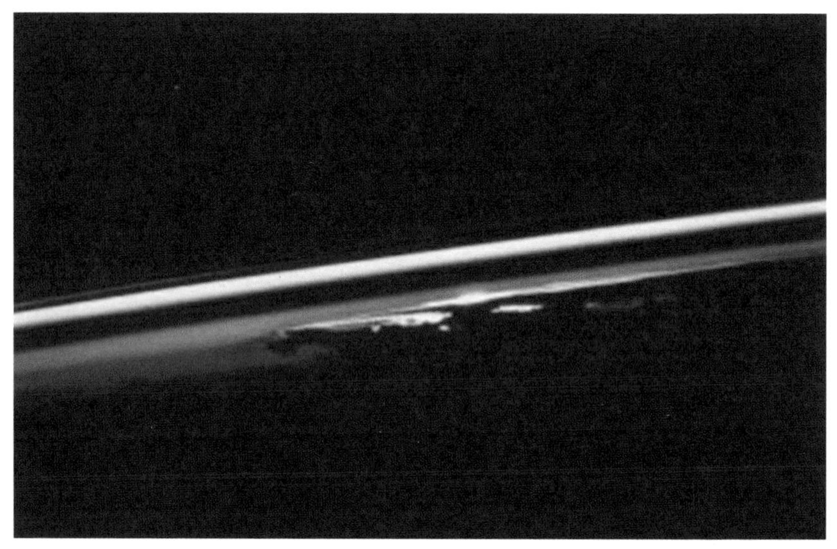

Figure 19-3: Sunrise from space, taken by STS-47 crew (NASA photo)

Mamoru had taken a lot of Japanese food with him to use in an educational video. The payload team had eaten these foods while training in Japan, but the orbiter team had not. One of the funniest things I remember from our flight is Curt eating a pickled plum, "umeboshi." His puckered face and watering eyes after eating the umeboshi were hilarious and provided some comic relief for the rest of us!

I had my own Japanese food on board—curry rice. In Japan, they had white rice pressed into a square about five inches by five inches and sealed in plastic. Also, the curry that was my favorite meal in Japan could be obtained in foil packages. It was perfect for spaceflight, and I'd brought some extras along for my crewmates. I'd also taken some fancy Japanese chopsticks with me, and with them, I broke off a chunk of the rice and dipped it into the curry (Figure 19-4). *Oishii*! It was more delicious than I remembered it being, and I think it was because my taste buds had changed. They really liked the spicy curry!

Before my spaceflight, I'd jotted down some things I wanted to do and pictures I wanted to take. At the top of that list in my crew notebook was to take a video of me twirling on my ice skates in space! One night while I was sleeping, Mark made some ice skating blades for me out of one of the stiff plastic covers on our procedures books and covered them with gray metallic duct tape. I taped these blades onto my muk luk socks (which I wore all the time) to make my ice skates and was videotaped "ice skating." I wanted to perform the world's first quintuple axel jump, which was five and a half turns, on ice skates. At that time in 1992, ice skaters could do triple axels but not quadruple axels. I knew it was a matter of time before quadruple axels could be done (and now they have), but I was pretty sure that a quintuple axel would never be done. So, I performed the first quintuple axel, and it is on video for proof and for posterity!

Figure 19-4: Jan eating Japanese curry rice in space (NASA photo)

In addition to requisite pictures of bumper stickers, pennants, and the crew, I wanted to take a photo of me eating Girl Scout cookies. As you read earlier, at that time I was a Girl Scout troop leader, in addition to being a Girl Scout myself, and I wanted to fly some Girl Scout cookies into space with me. I thought that the shortbread cookies in the shape of a trefoil would be the most easily recognizable, so the food laboratory folks had sealed some for me to take with me. The picture that was taken of me eating the cookies has become very popular (Figure 19-5), and I was happy to promote Girl Scouting, especially from space!

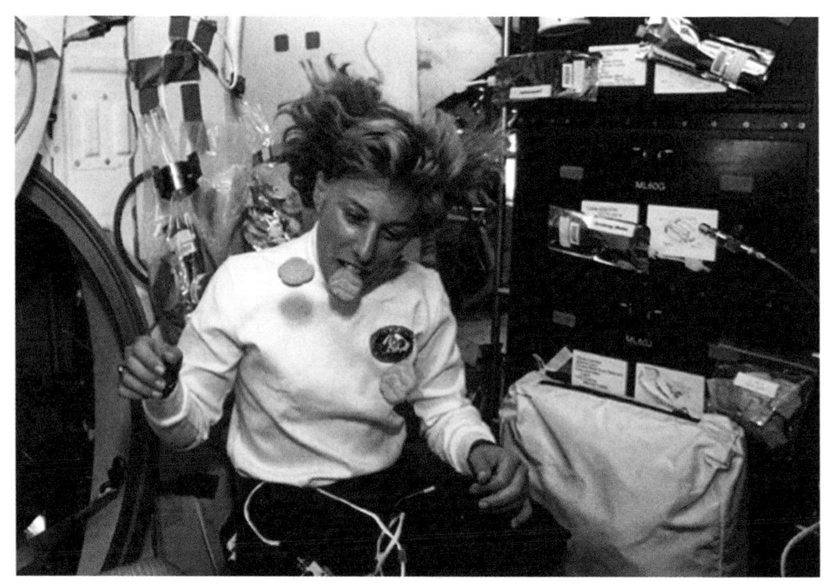

Figure 19-5: Jan eating Girl Scout cookies in space (NASA photo)

There were too many experiments conducted on our flight to go into detail about each one, but overall, it was a very successful flight, and the scientists were delighted with the results. I found the Spacelab module to be a highly effective laboratory with excellent facilities, and it worked almost flawlessly with its own climate control, vacuum and water loops (for cooling), data collection and transmission, television and recording capabilities, and so much more. It was a great example of microgravity experiments that could be done in a laboratory, and it paved the way for the laboratory modules that are now on orbit as part of the International Space Station (ISS). Our official crew picture in space is shown in Figure 19-6.

On my last day on orbit after we buttoned up the Spacelab, I had some extra time to gaze out the window for some last looks at the Earth. We had no more film for the Hasselblad camera that was used to photograph the Earth, so the only camera available for Earth viewing was the Linhof. As we were going over North America, I recognized the Great Lakes, and then I saw one of the most amazing things I ever observed from space. There was a huge brown dust pall over Canada that stretched for miles and miles. It looked so out of the ordinary, and I did not know what had caused it, but I knew it must be significant. I grabbed the Linhof and awkwardly took some pictures. This was difficult because the very large Linhof camera had no viewfinder, and it was hard to position it against a window.

Figure 19-6: STS-47 crew in space in the Spacelab J module (NASA photo)

I later found out that the dust pall had drifted to this area from a volcanic eruption of Mount Spurr in Alaska on June 27, 1992. The dust and ash from this volcanic eruption caused air travel to be rerouted around the dust pall (Figure 19-7). The earth scientists were very excited to get this photograph, so I'm glad I went to the trouble to take it. It was impressive to me because it made me realize how something in the environment in one part of the world vastly affects other parts of the world. Although this was a natural phenomenon, it demonstrated to me how we have to think about our entire Earth as our home and be careful how we treat it. Even if something we do may not affect us in our locale, it could affect the environment at a different place in another part of the world.

Prior to reentry, we deactivated a lot of equipment and took down things that were for on-orbit use only. All the cameras, samples, film, and anything that was loose had to be stowed in lockers or tied down in the airlock. The seats for launch and entry were reinstalled, and it was time to come home.

Figure 19-7: Astonishing dust pall over eastern Canada (NASA photo)

About an hour before entry, we were each to drink two quarts of water and sixteen salt tablets. Yuck! As I mentioned earlier, our bodies lost a lot of water while we were on our space journey (I lost seven pounds), and fluids in our body dwelled in our trunk area. Therefore, when gravity took over, the fluids would shift into the rest of our body and leave our heads. This phenomenon made some astronauts dizzy or lightheaded and even caused some to pass out once they reached Earth. I was determined not to have that happen, and I diligently drank my disgusting cocktail so the water (with the help of the salt) would be retained in my upper body and, more importantly, my head.

We retrieved our LESs and started the process of putting them on, along with our undergarments, basically repeating the process of when we'd donned our suits for launch. However, we did not have the suit technicians to help us this time! Even though we had been in space only eight days, our spines had lengthened (without gravity to compress them), and we'd each grown an estimated inch or two. That doesn't sound like a big deal, except that it was *painful* squishing ourselves back into the suits. I didn't expect that! Ouch! I will have to say it was easier to put the suits on in zero-g than in 1-g. Yes, Mom, I can put two pant legs on at the same time! That made things much easier.

I was excited about being able to experience reentry from the flight deck, as Mark and I exchanged positions. I had been in a lot of entry simulations with Hoot, Curt, and Jay, and I was looking forward to putting this knowledge to good use. We strapped ourselves in—Hoot first in the left seat, then Curt in the right seat, followed by me sitting behind Curt, and Jay sitting between Hoot and Curt.

About forty-five minutes before entry, it was a very busy timeline, as we turned our dear Endeavour around 180 degrees so it was tail first in the direction of flight. As we burned the orbital maneuvering engines, we could feel them lurch, and it was a bittersweet feeling to know we would not experience weightlessness any more on this flight. This burn would slow us down and cause us to go down, down, down into the Earth's atmosphere. We were only 190 miles above Earth during our mission, and it did not take a very long burn to start us in a downward direction. After the burn, we turned the orbiter back around, and it was doing its duty to glide us safely home. The nose was pitched slightly high so that the black tiles on the underside of Endeavour could protect us from the heat of reentry.

It wasn't long before we started descending through the atmosphere and experiencing quite a bit of shaking and buffeting. Most surprising to me as I looked out the overhead window was the bright orangey red glow I saw. It looked like the orbiter was on fire! The plasma generated by the heat and friction of going through the atmosphere went around the orbiter and reformed into a circle right above me. I had been briefed about this before, so it did not alarm me, and I viewed it as best I could with my wrist mirror. There were flashes of white and varying colors of orange, red, and yellow. My large helmet prevented me from turning my head upward to view the light show, but it was breathtaking to view it with my mirror.

After that, we began to feel the g's. From talking to other crewmembers and hearing medical debriefs from crews who had flown into space, we knew it was recommended to move our head left and right, forward and aft, and to basically be a bobblehead doll. Also, it was recommended to strain as much as we could against the straps, trying to use our muscles to "stand up" against them. This being my first flight, I thought I had better do those things, so I was moving my head around, bending forward, and standing up as much as I could.

Later, we did a series of S turns (called roll reversals) to increase our drag and slow our descent. This was enthralling for me to observe, and I was so impressed with the skills of our commander and pilot. We were going pretty darn fast, slowing down as we came closer to landing at KSC. Hoot and Curt were calling out Mach speeds and telling us the sights that they could see out of their windows. From the "back seats," we couldn't really see much of anything, so we appreciated the guided tour.

Everything happened so quickly, and before I knew it, we were nosed over, and I could see Florida. Soon, we were making the turn that is part of the Heading Alignment Cone (HAC). Basically, the orbiter flies in a pattern that is conical shaped while falling like a brick. There was quite a bit of buffeting now, as we were in Earth's atmosphere, where all the orbiter flying surfaces could make it fly like an airplane. However, we were not powered by engines, so the orbiter was actually a glider.

On the ground, bleachers had been set up for family and friends as they excitedly looked up into the sky to find Endeavour. She was still too high and too far away to see, but the loudspeakers quelled their fears, as everything was apparently nominal. Their first indication that Endeavour was almost home was the sound of two sonic booms as the orbiter approached.

As the HAC diameter grew smaller, Hoot began flying the orbiter and took it off autopilot. He flew Endeavour by managing the energy, altitude, and direction with his joystick—the rotational hand controller. We could all feel the turns and knew that we were almost home on planet Earth. I was recording the events with a camcorder as best I could, even with butterflies in my stomach against my tight g-suit.

Hoot and Curt could see the projected flight path with their heads-up display. I could see it and could tell that they were on track to make a perfect landing on the SLF runway at KSC. Turning onto the final approach, I felt myself press forward in my seat as we were flying on a nineteen-degree glide slope and very fast—so different from any of my previous aviation experiences. About three thousand feet above the ground, Curt had to perform one of his most important duties—to lower the gear. I stopped recording and put away the camcorder, as I did not want to damage it during the landing.

As Curt called out the airspeed and altitude, we could feel the thud of the gear being lowered and were glad that all was well. Hoot greased the landing—I honestly could barely feel it. He held the nose up as Curt read off the rotation rates so that the nose gear would land easily and avoid excessive slapdown loads.

One of the changes to the Space Shuttle after the Challenger accident was to add a drag chute that would be deployed after landing. Curt pushed the button to deploy the chute. Endeavour whirred while it decelerated, and the nose gear hit the pavement. We could feel our bird slowing down. It had been a good ship.

Everyone knew that Hoot was a top-notch pilot, and his landing of Endeavour that day proved it once again. As airspeeds were called out by Curt, Hoot slowly applied the brakes. When we slowed down, Curt jettisoned the drag chute with the push of a button. If he hadn't, the chute could have been tangled up in the SSME engine bells. After that was just one more gentle squeeze on the brakes.

Everyone on the crew excitedly congratulated Hoot on a super landing. After taking a deep breath, he very calmly radioed, "Endeavour, Houston. Wheels stop." Our almost eight-day adventure was over on September 20, 1992.

Outside, everyone in the stands was cheering for our safe return to Earth. During WWII, the ground crews and colleagues of those who had flown that day in combat had watched the B-17s with anxiety as they'd returned to home base. They'd counted the airplanes and noted the ones that had not made it back. The landing of Space Shuttle Endeavour was a little different than that, but anxious families, friends, and colleagues also watched with hopeful anticipation that the crew would come home safely.

After getting up and walking around, I did not have problems with orthostatic intolerance, dizziness, lightheadedness, or any other maladies that sometimes affect astronauts after landing back on Earth. I was able to join the rest of my crewmates in doing a walkaround of the orbiter, euphemistically "kicking the tires," and examining the tiles. She was a good ship who had taken good care of us for eight days.

Each crewmember had an "assistant" who was supposed to support us if we felt woozy or in danger of falling. I did fine until I looked up at the tiles while walking underneath Endeavour. Whoa! I felt like I was going to fall over backwards. After my assistant grabbed my elbow to help me, I was able to maintain my balance without embarrassing myself on national television.

After taking a final look at Endeavour, we were bused back to crew quarters at KSC. We took showers (that felt so good) and changed clothes and maybe had a bite to eat. Now I knew what the POWs had gone through when they could not have regular hot showers! Then, it was time to see our anxious families. Eight days had gone by fast for me, but I think it had gone by slowly for the families on the ground who were concerned for our safety.

We had a happy reunion with our families and enjoyed a very nice visit before we had to fly back to Houston. The homecoming at Ellington Field, our home base, was such a memorable one. Many of my friends, coworkers, instruction team, and fellow astronauts heartily greeted us with smiles and hugs. My Girl Scout troop, for which I was a co-leader, were all there! They'd made a big sign saying, "Way to Go, Jan," and carried it proudly. It was so good to see everyone!

That night, when I was in my own bed, I fell off it a couple of times. I guess after you have been floating for a week, your body doesn't realize you are amid gravity again. Apparently, I thought I was floating and just rolled right off the bed! I was not hurt, only amused by the experience.

There was no rest for the weary. The very next day (and for the next two days), I had to do more baseline data collection. When I was not doing that, I reviewed mission photographs with the rest of the crew. It was critical to release twenty-five photographs for the public affairs office, and we also used some of the photographs in our debriefings.

As we returned to our office in Building 4 at JSC, we discovered our instructor team had decorated the halls with funny cartoons and quotes that we had experienced in training or in flight. We'd had a lot of fun during our training, so there was a lot of material! Our training team had done an awesome job getting us ready for flight, and we were very close to them. It meant so much to us for them to decorate the halls of our building and our offices. What a fantastic welcome home!

After that, there were three weeks of debriefings for every group who had an interest or involvement in our mission. We briefed our Flight Operations Directorate, the flight directors in the Mission Operations Directorate, the Astronaut Office, and the JSC senior staff before we had our post-flight U.S. press conference and a Japanese press conference on September 29. Then, we had more debriefings, including a trip to MSFC to debrief the Spacelab J hardware and software, training, and payload operations.

At some point in the middle of all those debriefings, we had a splashdown party in Houston! It was great to see all our friends, training team, flight controllers, and others who had worked so hard to make our mission a success. It was our way of thanking them for a job well done.

As soon as the debriefs in the U.S. were completed, in early October 1992, I was honored by my hometown of Huntsville with a parade and events all over town. It was called "Golden Homecoming," as Spacelab J was the fiftieth flight of the Space Shuttle. It was quite a celebration since I was Huntsville's own astronaut and MSFC played a key role in not only the Space Shuttle propulsion but also the management of Spacelab missions such as ours. I had nine appearances that day, and each one was impressive with big gold bows, balloons, bands, dancers, lots of children, and the ever-eager public. It was a day I will never forget.

After a string of debriefings and celebrations, we had a little time off before we went to Japan in November 1992 for even more debriefs. The country of Japan was ecstatic about the successful mission, and they were so proud of their astronaut Mamoru, so I knew it would be a great trip.

In addition to technical and training debriefs, we attended several celebrations hosted by different groups in Japan. We traveled all over the country on our post-flight visit—to Tokyo, Kobe, Tachikawa, and Tsukuba. We were wined and dined everywhere and met many dignitaries. Plus, we were featured speakers at the Pacific International Aerospace Foundation conference.

My favorite place to visit on that trip was Kobe. We were treated as guests by the Japanese at a ryokan, which is a traditional Japanese guest house. Each of us had a private room, complete with a tatami mat, a futon mattress, and tables low to the ground so that we had to squat to be seated. We were waited on by beautiful and gracious Japanese women, all dressed up in their kimonos, who served delicious and exotic food.

As Kobe is known for its hot springs, the ryokan where we stayed had a traditional hot bath, or onsen. Men and women were separated, and bathrobes (yukata), towels, and slippers were provided in our rooms. Taking a traditional Japanese hot bath was a fantastic cultural experience, and I have never felt as relaxed as I did after taking that hot bath in Japan.

After the unforgettable trip to Japan, we had some breaks in our schedule, and we celebrated Thanksgiving and Christmas in ways that we had not been able to for a few years. It was nice to take some time off with our friends and families, but it was also kind of a letdown not to spend more time with the rest of my crew. They had become close friends, and I missed them. They are dear friends of mine to this day, and I always enjoy our astronaut reunions and any occasion when I can see them again.

As mentioned previously, my father had a similar camaraderie with not only his flight crews and the classes with whom he trained but also with the POWs who were imprisoned with him. Their training classes included close friends who went through good times and bad to get their wings and their ultimate aircraft assignment. Then, when the guys went overseas to their respective bases and squadrons, the environment of combat led to a special bond that is shared like no other. The POWs relied on each other for physical and mental health, and through stressful situations, they took care of each other. On-orbit in the harsh environment of space while adhering to a pressure-packed schedule, that is what our Shuttle crew did—we worked well together, and we took care of each other.

At the time of our flight, we did not know whether Chiaki or Takao, our backup Japanese astronauts, would ever fly in space. However, they both had opportunities to fly on the Space Shuttle, and Mamoru Mohri flew one more time as well.

Due to my first spaceflight taking so long (five years) to come to fruition, the Astronaut Office hierarchy wanted to assign me to another spaceflight mission as quickly as possible. Little did I know that in late 1992, I was going to be assigned to another important international mission!

Chapter Twenty
Flying with a Russian

Our country's space program was firmly entrenched in the Space Shuttle program, and NASA had struggled to gain approval from the Senate and Congress to build a space station. Going to space is expensive, and the only way to make a space station affordable for the U.S. was to involve some international partners. International partnerships had been successful for some Space Shuttle flights, like my first flight, which had been a partnership with Japan. The respective state departments of various nations and the legal entwinements of cooperating with other countries had been worked out with the Space Shuttle missions, so it was time to kick it up a notch.

On Christmas Day in 1991, the Union of Soviet Socialist Republics (USSR) dissolved, and President Mikhail Gorbachev resigned. NASA took the opportunity of the changes in Europe and Russia to develop a partnership that would be a win-win for everyone. ESA is a consortium of many European countries with varying levels of participation and financial commitment. By partnering with ESA and Russia, a space station could be affordable for these countries and the United States.

Russia had extensive experience with flying long-duration missions on a space station during the Salyut and Mir programs. On the other hand, the U.S. had extensive experience with a reusable and successful Space Shuttle. In 1992, NASA executive George Abbey made a brilliant idea come alive when he announced that the United States and Russia had signed a new agreement under which they would partner for cooperation on space exploration.

Phase One of the joint Russia-United States cooperation was called the Shuttle-Mir program. It would involve flying Russian cosmonauts on the Space Shuttle, flying American astronauts on the Mir Russian space station, and docking the Space Shuttle with the Mir. If Phase One was successful, then it would pave the way for the Phase Two construction of an international space station.

The first Phase One mission of the Shuttle-Mir program was a Space Shuttle mission that would include a Russian cosmonaut as a member of the flight crew. In December 1992, I was ecstatic to be assigned to this mission as a mission specialist! Not only did I realize the historic nature of the mission, but I also was really looking forward to learning more about Russia, its space program, its culture, and the cosmonaut with whom I would fly.

The mission was designated STS-60, and the assigned crew was led by veteran astronaut Charlie Bolden as the commander and included my classmate Ken Reightler as the pilot. Franklin Chang-Diaz was the payload commander, and rookie Ron Sega rounded out the American crew. Veteran cosmonauts Sergei Krikalev and Vladimir Titov were the Russian cosmonauts on our crew with one chosen as prime and one chosen as backup in 1993. It was a great crew, and I was excited to be a part of it!

The first order of business was to make our Russian cosmonauts feel at home in the environs of JSC and Houston. In fact, there was an entire office at JSC devoted to handling the logistics of welcoming the Russians to Texas. They took care of things such as renting vehicles to drive; renting a house; hiring language instructors to teach English; and setting up bank accounts, insurance, badging, and anything else that was needed.

As a crew, we tried to explain and demonstrate our culture, our shopping places, our customs, and our slang language. As Charlie said, "In getting them into the communities, getting them to do the kinds of things we did, happy hour, church if they wanted to, but those kinds of things, we just kind of took them by the hand and let them see and do as much as they wanted to." With that in mind, I made it a point to take my crew to Huntsville so they could get some good southern food and grits at Eunice's Country Kitchen!

Sergei brought his wife, Elena, and his precious three-year-old daughter, Olga, with him. He spoke good English and came from the civilian side of the Russian cosmonaut corps. A champion aerobatic pilot and an avid competitive swimmer, he'd had more time in space than the rest of thc crew combined! Sergei was a mechanical engineer from St. Petersburg who had been a flight engineer on his first spaceflight, which had been a mission to the Mir space station in 1988 and 1989, and he'd stayed in space 151 days. A soft-spoken, mild-mannered man, Sergei was very smart and was a quick study in learning about life in America as well as life on the Space Shuttle. At the time of our spaceflight mission, Sergei had the world record for the longest time in space.

Vladimir Titov, whom we affectionately knew as "Volodya," was the other cosmonaut assigned to our mission. He was a happy, jolly sort of fellow and was always smiling and ready to laugh. Volodya came from the military side of the Russian cosmonaut corps and was a pilot and colonel in the Russian Air Force. He was also an experienced cosmonaut, having flown a Soyuz spacecraft to the Salyut space station in 1983. He then flew another Soyuz to the Mir space station and lived there in 1987 and 1988. On his spaceflight aboard the Mir space station, he and his crewmate had spent 366 days in space, setting a new spaceflight endurance record.

Volodya's most exciting spaceflight experience had been on board a Soyuz spacecraft that was supposed to launch in September 1983. However, a valve in the propellant line had failed to close at T-90 seconds, causing a large fire at the base of the launch vehicle only one minute before launch. The fire had quickly engulfed the rocket, and launch controllers had manually aborted the mission twelve seconds after the fire had begun. The launch escape system pulled the Soyuz module away from the launch pad, and after being subjected to fifteen to seventeen g's, the crew had landed safely 2.5 miles from the launch vehicle, which had exploded seconds after the Soyuz separated.

I always enjoyed hearing Volodya's version of the launch abort, as he used his broken English and hand gestures to describe what happened. "We went poof!" he'd say with one hand atop the other, and then his second hand would fly off the first and land far away!

Volodya brought his wife, Alexandra, whom we affectionately called "Sasha," with him, along with their two children, an eight-year-old son who enrolled in public school in Friendswood and an eighteen-year-old daughter who spoke fluent English and enrolled in San Jacinto College in a pre-business curriculum. The more English that Volodya learned, the funnier he became! He had a great sense of humor and was able to express himself better when he became more comfortable in English.

It was a marvel to our crew, especially the two military men, Ken and Charlie, that we had such a good relationship with the Russians. As Charlie once said, "We were trained to kill each other," and that was a fact.

Charlie Bolden was a Marine colonel and pilot from Columbia, South Carolina, who is just a super guy. He grew up during the rough civil rights history of South Carolina, and he could not get a military academy appointment from his own state because he was African American. As a high school student, Charlie wrote a letter to then President Johnson, asking for an appointment to the U.S. Naval Academy. He later received this appointment from an Illinois congressman. He'd had three prior Space Shuttle missions, and our STS-60 flight would be Charlie's last. Although it was not the mission that he had hoped for (he wanted to be commander of a Hubble repair mission), he really enjoyed all aspects of the flight, the training, the crew, and the instructor team.

Charlie is a strong family man, and he wanted to include the Russians' families in all our social activities, along with the family members of the crew. We had a social event at least once a week, where everyone could get together, including all the children. Most of our social activities also involved the training team, as Charlie noted, "We had a tremendous training team. The training team, as always happens, became very close to the crew because they live with them, eat with them, sleep with them. You're just with them all the time. So, they became very close." In fact, throughout the training and flight, we all became like family. I am still close to the STS-60 crew and to the training team.

Ken Reightler was a Navy pilot from Virginia who had been in my astronaut class. While we were ASCANS, I'd become close friends with Ken and his wife, Maureen, as well as their two daughters. When Ken and I were assigned to STS-60, it was to be the second flight for both of us. He told me, "There is no one who I would rather fly with," and I felt the same way.

Our payload commander, Franklin Chang-Diaz, was a very interesting character with an incredibly diverse and amazing background. He grew up in Costa Rica, the son of a Costa Rican mother and a Chinese worker. After moving to the United States when he could barely speak English, he attained his bachelor's degree in mechanical engineering from the University of Connecticut before earning a Ph.D. at the Massachusetts Institute of Technology.

As we were assigned to our mission with only a year to train, we were on a fast track. The Spacehab module was about one-third the size of the Spacelab module that had been on my first mission. It did not have nearly the capability of the Spacelab module, but it was a smaller, less expensive way to augment the Space Shuttle middeck for various experiments. Franklin and I were responsible for the Spacehab portion of the mission, and this was the first flight of our module (Ernie) and the second flight of a Spacehab (the first module was named Bert). There were twelve Spacehab-sponsored experiments, including four from my alma mater, UAH.

There were also three LS experiments, sponsored by the Russian Institute of Biomedical Research in Moscow and JSC's Space and Life Sciences Directorate, and a joint Russian-U.S. Earth observation study. Filling out the manifest were some less complex Getaway Special experiments in the payload bay.

The other main payload on our flight was the Wake Shield Facility (WSF). It was a twelve-foot diameter dish that looked like a flying saucer and was designed to take advantage of the high vacuum of the space environment to produce new materials for use in semiconductors and other electronic devices. Ron Sega, the rookie crewmember, was the principal investigator (PI) of the WSF. Very professorial in nature and an expert in the WSF, Ron was a fun and delightful person with whom to train.

In April 1993, Russia decided that Sergei would be the prime cosmonaut on our flight and Volodya would be his backup. We continued training with both for on-orbit operations, but the ascent and entry simulations were only with Sergei.

I was beyond excited when Charlie decided that I would be the prime operator for the CANADARM Remote Manipulator System (RMS) to deploy and retrieve the WSF and that Sergei would be my backup. With Franklin as the payload commander with prime responsibility for Spacehab and its experiments, I would be the backup for him. Other crew assignments I had were EVA duties with Franklin and crew medic duties with Charlie. I was plenty busy with training! I loved every single minute of it.

For my father's B-17 crew of ten people, the division of responsibilities was very similar to that of a Space Shuttle crewmember. Each crewmember on the B-17 had a specialty for which he was trained but also had a backup. Everyone had to train on everything to some degree, but the specialists homed in on training for their area of expertise. They had to train on gunnery, life support (oxygen), radio and communication procedures, crew escape and emergency procedures, donning and doffing equipment, and much more. The pilot, copilot, bombardier, and navigator also had to train on geography, flight mechanics, aircraft systems, navigation equipment, bombing procedures, malfunction procedures, security, and evasion techniques, in addition to their specialties. There was a lot of cross-training involved, just as there was on the Space Shuttle. The pilot of a B-17 and the commander of a Space Shuttle mission had many things in common—they were responsible for their crew and their mission and were the leaders who had to make critical decisions that could make a difference in life or death.

In 1993, it was announced that Russia and the U.S. would cooperate on Phase Two of the Shuttle-Mir agreement, the construction of the ISS. ESA, Canada, and Japan were also solicited for their participation, but Russia and the U.S. would lead the development and construction. In addition to Russia, ESA, Canada, and Japan would also provide hardware on a barter system in exchange for flying astronauts from their countries.

Training for STS-60 was intense and thoroughly enjoyable as the crew melded and worked together. Charlie did an amazing job of integrating the families, the instructors, and the crew. By the time we were going to launch in late 1993, we were a highly functioning team. In reviewing my STS-60 notes, it was mind-boggling how many details were covered, in addition to our Space Shuttle launch, entry, and rendezvous training.

As was typical for Space Shuttle flights, we were quarantined one week prior to launch and spent the first five days of that in the JSC crew quarters. Charlie invited his Episcopalian priest to serve us communion while we were in quarantine, and that was a very special time for me. I appreciated

Charlie being a role model in living out his Christian faith and making it a part of his important position.

Our flight was delayed to early 1994, as often happens with space flights or rocket launches. On February 3, 1994, in the morning hours at KSC, we launched on Space Shuttle Discovery—on time at the opening of the launch window. Once again, we were lucky! For this launch, I was on the flight deck, sitting behind Ken in the right seat. Ron was the flight engineer to my left, and Sergei and Franklin sat on the middeck. The countdown to launch went by fast, as things were busy on the flight deck while we prepared for launch. I was relaxed and eager to get to space.

The countdown went smoothly, and the launch was easier for me the second time. The only surprise was how big the flash was when the SRBs separated from the vehicle.

Figure 20-1: A photo of the external tank after being jettisoned from the Space Shuttle (from STS-114, NASA photo)

One of my main duties was to photograph the ET after it was jettisoned. After eight and a half minutes, the SSMEs shut down, having used up all the fuel that was in the ET. I awkwardly climbed out of my seat, as I was wearing the bulky orange LES, and unstowed the camera and long lens from the flight deck locker. Balancing on top of one of the seats, I photographed the ET as it rolled away (Figure 20-1), later breaking up in the atmosphere over the Indian Ocean. The ET was the only part of the vehicle that was not retrieved, and it was important to get good pictures of it so that analyses of any potential problems could occur.

The view of the Earth was again spectacular, and it is difficult to describe the sensation of floating and seeing our Earth from space. Now, it seems like it was all a dream, but I will never forget that view. The winter season in the northern hemisphere afforded us some spectacular views of Russia, and we were able to take some very nice photographs of Sergei's hometown of St. Petersburg. Of course, in the 2020s, we can readily get pictures of the Earth from satellite imagery, but I still like to look at the photographs that our crew took as we passed over our beautiful Earth every ninety minutes.

Figure 20-2: A photo of Kennedy Space Center and Canaveral AFS, taken from the Space Shuttle, STS-60 (NASA photo)

One of my favorite photographs taken on STS-60 is a picture of KSC from where we launched (Figure 20-2). In this photo, you can clearly see the SLF's long runway as well as the shorter runway, the Skid Strip, at the Canaveral AFS. It is located near the "point" of Cape Canaveral and is very close to where my father tested and operated Matador missiles in the 1950s and where I was born at nearby PAFB. Canaveral AFS and PAFB are now part of the U.S. Space Force military branch. No doubt, if my father were alive and on active duty today, he would be a member of the U.S. Space Force.

I was able to look at the Earth more often on this flight than on my first flight, and I pressed my face to the glass whenever I could. Having Sergei on the flight was like having a real-time talking map, as he knew all the features and cities on the Earth due to his long duration in space. He was so comfortable in space and taught us lots of tricks on how to have fun with the microgravity environment. The aurora australis (in the southern hemisphere) was particularly active at that time and was even more beautiful than what I'd observed during my first spaceflight mission.

After we took our orange pumpkin suits off, it was time to get to business! On that first flight day, Franklin and I flew down the tunnel to the Spacehab module, opened the door, and brought the module to life for the first time! However, when we connected the air hose from the Space Shuttle to the Spacehab, the hose collapsed. We told mission control about it, and they didn't seem too concerned, so we continued activating Spacehab and its experiments, and all went well.

On this mission, we were single shift, so all of us went to bed at the same time. We crawled into our individual sleeping bags and "hung out" wherever we could. Although I was tired, it was hard to sleep that night because the next morning I would be checking out the CANADARM and operating it for the first time.

The following morning, mission control played "Sweet Home Alabama" for me as my wakeup music. I responded, "*Dobroye utro*, y'all!" ("good morning" in Russian). It was thrilling to "fly" the CANADARM, and my first experience with it was to do a payload bay survey and get the feel of operating the arm. Sergei and I took turns, and we felt the training simulators had done a good job in getting us ready.

Later, we unberthed the WSF, and the ground controllers began their checkout. Unfortunately, due to a faulty horizon sensor on the WSF, I was unable to deploy the WSF, and we had to leave it attached to the arm (Figure 20-3). We did various maneuvers with the WSF on the arm, and the scientists were able to grow some thin film crystals, even though the environment was not what they'd hoped. Fortunately, the WSF was able to be deployed on a later Space Shuttle flight.

Figure 20-3: Wake Shield Facility on the CANADARM Remote Manipulator System (RMS), STS-60 (NASA photo)

Even more disappointing for Charlie and Ken, they would not be able to perform a rendezvous of the Space Shuttle and the WSF. During a rendezvous of the Space Shuttle and a free-flying object, the pilot and commander had to conduct a series of OMS burns and RCS (small thruster) burns, and they would "fly" the Space Shuttle to the object. A lot of planning and careful piloting of the Shuttle to perform the orbital mechanics was necessary. So, it was a highlight of a mission for a pilot and commander to do a rendezvous.

Although I was also disappointed that I would not be able to "fly" the arm to capture the WSF after the rendezvous, I was happy to be able to operate the CANADARM as much as I did. Of course, as the PI for WSF, Ron was even more disappointed than the rest of us, so we tried our best to cheer him up.

Now, I need to tell you the story about the collapsed hose. Franklin had a live video event with Costa Rica and was giving them a tour of the Space Shuttle and the Spacehab. As part of that, mission control observed the collapsed hose and told us that we needed to fix it because they had not realized how bad it was from our previous description.

By this time, Sergei was frustrated with the constant communication with our ground team—the Tracking and Data Relay Satellite System gave us complete communication coverage around the world, except for a few minutes each orbit. For the Russian spaceflight missions, they'd only had voice communications while flying over Russia; therefore, the Russian cosmonauts were accustomed to having a lot more autonomy.

While we were waiting for instructions from mission control about how to fix the collapsed hose, Sergei said, "Do what we do in Russia—the crew fixes something and then shows the ground what they did." So, that's what we did! We took a hard plastic cover from one of our map books and rolled it into a tube, securing it with duct tape. During a loss of signal (LOS) from mission control, we inserted the tube into the collapsed hose, and this gave it enough stiffness that it would not collapse. When we returned to communications with mission control after the LOS, we relayed our innovative solution to them by downlinking a camcorder video of our handiwork. After a few minutes of silence, they responded, "Discovery, Houston—that looks good." I think this was an important lesson for our space program that we learned from the Russians, and hopefully it has continued with the joint Russian-U.S. operations on ISS.

We had all worked very hard to practice our Russian for speaking with the Russian press during our mission. Yuri Koptev, the head of the Russian Space Agency, had visited mission control in Houston, and we spoke with him while we were in space. In addition, we spoke with President

Bill Clinton and Russian Prime Minister Boris Yeltsin. President Clinton commented that he liked my hair floating in space!

The highlight of our Russian-speaking events was a song that we had prepared to sing to the Russian children. It was commonplace in Russian television for a song "Tired Toys Are Sleeping," to be played, signifying to the children it was time to go to bed. We sang this song live to Russia at the proper time, and hopefully, we sent a nation of children off to sleep!

I am often asked what my favorite experiment was on my spaceflights. Without a doubt, it was the Protein Crystal Growth experiment on STS-60. We were carrying two batches of a solution of the protein insulin. The purpose of the experiment was to grow a single and pure crystal of insulin so that scientists could determine the exact structure of insulin post-flight. The advantage of growing crystals in space is that their growth is unencumbered by the container or distortions that occur when growing a crystal in the gravity of Earth. As I monitored the experiment, the crystal started growing right before I was supposed to go to bed! Mission control wanted me to go to sleep, but I refused, as I had to change some experiment parameters to keep the crystal growing (Figure 20-4). I was delighted to learn after the mission that we had grown a perfect crystal of insulin and that scientists could then know what the exact structure was. This discovery helped in the research and treatment of diabetes and the frequency needed for insulin injections.

Figure 20-4: Jan operating Protein Crystal Growth experiment on STS-60 (NASA photo)

Sergei was an avid amateur radio user, and we were able to get the Shuttle Amateur Radio Experiment (SAREX) manifested onto our flight. Therefore, Sergei was able to hook up to his three friends and colleagues on the Russian Mir crew by radio and have conversations with them. I was also able to use SAREX to connect and speak with my mother's students as well as talk with students from my grade school in Huntsville. It was a great asset to the flight, as we could converse freely with people on the ground and expand children's interest and involvement in space.

We tried to incorporate as many Russian items as we could on our flight. We took some Russian space food, and I thought that it was very good. Some of the Russian delicacies continued to be used on the Space Shuttle after our flight and later on the ISS. One of my favorites were little loaves of bread, which were delicious, as we typically couldn't eat bread in space due to the crumbs going everywhere (Figure 20-5). We took a Russian flag, Russian science experiments, and a small toy Cheburashka (Russian cartoon character) to serve as a "zero-g indicator."

During Russian spaceflights, each cosmonaut carried a small stuffed creature tied to their kneeboard. After experiencing the g's during launch, the zero-g indicator would let the cosmonaut know when they were in space. This practice was adopted by American astronauts and was used on Space Shuttle flights, ISS flights, and commercial spaceflights. Recently, an astronaut stuffed Snoopy was used on the Artemis I mission as a zero-g indicator! A picture of our happy crew on orbit is shown in Figure 20-6.

After eight days on orbit, it was time to come home to planet Earth. I know that our mission was a historic one, and I was grateful to be a part of it and to make lifelong friends with Sergei and Volodya and their families. Fortunately, Volodya would get the opportunity to fly in space on the Space Shuttle, and Sergei would also have more opportunities to fly in the American space program—both on the Space Shuttle and in the ISS.

For entry, I was on the middeck, and Sergei was on the flight deck. It was uneventful, and Charlie and Ken kept Franklin and me informed about what was going on as we entered Earth's atmosphere and prepared to land. Once again, I drank the required salt water before entry, stood up as much as I could against the seat belt and shoulder straps, and swung my head from side to side. These techniques had seemed to work on my first spaceflight, so I continued the practice.

Figure 20-5: Sergei Krikalev and Jan eating Russian bread on middeck of Discovery, STS-60 (NASA photo)

After Charlie greased the landing at KSC on February 10, 1994, we were not allowed to stand up or get out of our seats. For medical research purposes related to the scientific experiments that we'd conducted on our flights, we were required to be put onto stretchers and taken out of the Space Shuttle. The vehicle outside of the Space Shuttle, much like the "people movers" at the airports, was waiting. We decided as a crew that everyone would depart the Space Shuttle in this way; therefore, no one was able to do the typical walk around the Space Shuttle after landing.

The flight surgeons and other medical personnel took all of us out on gurneys and moved us into the hospital-looking room to start collecting data. Yes, we were the guinea pigs who supplied blood, urine, and any other bodily products for the good of science. We were poked, prodded, measured, and checked out in almost every way possible.

I begrudgingly provided all the necessary data that the scientists wanted, and I was very pleasant about it. After all, this was how they made a living, and they were trying to do research that may help space travelers in the future. Fortunately, I had no problem standing up or moving around, and my post-flight results were the same as my first flight.

After visiting with friends and family at KSC, we flew to Houston and had a great reception at Ellington Field. This was followed by a good night's sleep! After the requisite debriefings over a few weeks, we began planning our trip to Russia. We wanted to make sure that we went there when it was warm, so we planned the visit for June 1994. Spouses were invited, and we had a warm welcome and a remarkable trip to Moscow, Star City (cosmonaut training and operations center), and Sergei's hometown of St. Petersburg.

Figure 20-6: STS-60 crew on board Space Shuttle Discovery. Clockwise from center: Jan, Franklin Chang-Diaz, Sergei Krikalev, Ken Reightler, Charlie Bolden, and Ron Sega (NASA photo)

It was interesting that, in Moscow, we stayed at a hotel that had been converted from the KGB (Komitet Gosudarstvennoy Bezopasnosti—the Russian Committee for State Security) headquarters. It was a beautiful hotel with dark, polished wooden floors; marble everywhere; and ornate chandeliers. There was a "host" or "hostess" on each floor who watched our comings and goings and was happy to be of assistance. We assumed that everything in the hotel was wired, so we were careful about what we said. We enjoyed fabulous tours, and we wined and dined through Moscow and Star City. Of course, there was a lot of vodka consumed and countless toasts to a great mission and to new partners in space.

After taking a train to St. Petersburg, we stayed at one of the nicest hotels there: the Astoria. We visited fabulous churches, palaces, and the State Hermitage Museum. Despite this opulence, my favorite part of visiting Sergei's hometown was eating at his parents' flat. His father was a shipbuilding engineer, and his mother was a teacher. I think they probably spent a month's wages on a feast for us. We had meats, cheeses, desserts, veggies, salads, and, of course, vodka. I wish that my Russian speaking skills were better so that we could have conversed more, but it was a heartwarming visit nonetheless.

The military members of our crew had to pinch themselves, as they'd never expected to view the crown jewels in the Kremlin or visit the country that was once our enemy. Likewise, I was pinching myself during my spaceflights and post-flight visits to Japan and Russia. Both spaceflights were significant to cooperation with other space-faring countries, and I feel fortunate to have been asked by NASA to be a part of building relationships with the space programs and people of these two places. The research, relationships, and beginning flights with these two partnering countries laid the foundation for future international relationships with Russia and Japan in constructing and operating the ISS. I could not imagine better missions or crews, and I wondered what my next spaceflight and crew would be like!

Chapter Twenty-One
Final Flight

My job after my second spaceflight was branch chief for the Mission Development Branch in the Astronaut Office. My branch was responsible for the development of procedures, training, and trade studies for EVAs, CANADARM operations, and payload operations. Our group performed the functions that an assigned crew would normally perform and provided evaluations to those who were developing new equipment or new approaches to operations for these three categories. It was a job I thoroughly enjoyed, and I had experience in all three areas. We were able to make decisions early in the development of hardware or processes, which was key to the success of a flight crew when it was eventually assigned to a mission involving these operations.

As I continued to work as Mission Development Branch chief and to go on public speaking engagements, I enjoyed having a little more time off than when I had been training for spaceflights or intensely working as a CAPCOM. I was able to take some quilting classes and become more involved in the local quilt guild in the Clear Lake area. Sue Garman, the associate director at JSC, and I developed an even closer relationship during this time, primarily due to our common interest in quilting. A quilt that I made from one of Sue's patterns is shown in Figure 21-1. I also had more time with my Girl Scout troop, which was always a pleasant relief from the technical aspects of my job.

I knew that my days as an astronaut were numbered, so I prepared myself for the next phase in my career. I enjoyed working with people while still directing technical projects, so I asked the chief of the Astronaut Office, Bob Cabana, for a position where I could increase my chances of one day being a NASA executive and entering the Senior Executive

Service (SES), which was the equivalent of becoming a military general. To do this, I needed training on how to be a supervisor, and I needed to learn the procurement and government budget processes. Therefore, I enrolled in NASA senior management classes around the country, and Bob appointed me as the head of a Source Evaluation Board (SEB) in mid-1995. In federal government procurement, the SEB selects the contractor who is awarded the contract after a long, competitive process. The year that I was head of the SEB was hard work, but I learned a lot, and I had a fabulous team who made it bearable.

Figure 21-1: Halloween quilt made by Jan from Sue Garman pattern; hand appliqued, machine pieced, and machine quilted (quilting not complete)

In September 1996, Bob Cabana assigned me as the payload commander and operator of the Japanese robotic arm on Shuttle mission STS-85. I was delighted to have Curt Brown as my commander, as we'd become close friends when he was in my 1987 astronaut class, and he'd been the Space Shuttle pilot on my first spaceflight mission in 1992. The rest of the crew included four rookie space flyers: pilot Jeff Ashby and mission specialists Bob "Beamer" Curbeam, Steve Robinson, and Canadian astronaut Bjarni Tryggvason. I could tell right away that this would be a great crew.

In addition to the Japanese robotic arm, the other primary payload on this mission was the German Cryogenic Infrared Spectrometers and Telescopes – Shuttle Pallet Satellite (CRISTA-SPAS). Having been in space once before on STS-66, the CRISTA-SPAS satellite was deployed and retrieved by the CANADARM RMS. The satellite was used to map the levels of ozone in the atmosphere at different altitudes, and this information was to be used to validate meteorological computer simulations of ozone in our atmosphere. Again, I was assigned to be the primary operator of the CANADARM RMS, and I was thrilled about that assignment.

As MS1 on the Space Shuttle Discovery, I would again be on the Space Shuttle flight deck for the launch and on the middeck for the entry phase of the flight. Therefore, I participated in all the ascent simulations in the SMS, along with Curt, Jeff, and Beamer. Beamer was MS2, the flight engineer, who sat in the middle behind the commander and pilot, and I sat next to Beamer. I took on the responsibility of knowing the information in the FDF procedures for the reference data, thus helping with failure signatures and malfunctions of bad sensors or missing data.

One of the improvements made since my last spaceflight was a new cooling system for our orange LESs. Upgraded to now be called the Advanced Crew Escape Suit (ACES), we wore cooling garments that consisted of a shirt and pants resembling thermal underwear with an extensive network of tubes woven into them. Water pumped through the tubes during the launch countdown kept the astronaut cool, and ventilation was also better than in the LES. A set of plugs folded into a pocket on the outside of the ACES connected to fittings inside the Space Shuttle (or simulator) to move the water through the suit. Ahh, instant cooling and much more comfortable than the LES. In fact, sometimes it was too cold, and we had to "dial it back a notch" to remain comfortable!

I don't think I have mentioned how we handled our "biological needs" during the several hours that we were in the LES or ACES. It was really a simple solution—we wore adult diapers. However, NASA had to have a fancy name for everything, so they were called Maximum Absorbency Garments!

As our training was underway in October 1996, Steve Robinson was assigned as my backup for the operation of the CANADARM RMS and the Japanese robotic arm. We learned that the latter was called the Manipulator Flight Demonstration (MFD) and was a flight test of the Small Fine Arm that was going to be flown on the exterior pallet of the Japanese module on the ISS.

Steve and I were scheduled to go to Japan in mid-November 1996 to train on the MFD. I was looking forward to being in Japan again, connecting with old friends, and making new friends there. The prime contractor for the MFD was Toshiba, with robotics work done by Hitachi. The integration of the vehicle was done by IHI, and we would see the flight hardware at the Toshiba plant near Tokyo and do some training there.

Our NASA robotics instructor was James Tinch (Tinch-san), and our NASDA point of contact for MFD was Yashio Kashiyama, whom we called "Sho" or "Loco." I thoroughly enjoyed being in Japan again, and the Japanese MFD technical team welcomed Steve and me with the typical Japanese hospitality. We only stayed in Japan a week, but we made the best of it and created great memories and friends.

MFD was going to be a challenge to operate in space, as it was a precision instrument that had some operations of the arm "flown" by ground controllers. I thought my biggest challenge would be operating the arm without being able to look at it out the window. Due to its position on the ISS, the MFD equivalent on the ISS would have to be observed with television cameras only. "Flying" the arm without visual contact was one of the things we would be testing on orbit.

It was so pleasant having four rookies on our mission. Curt and I enjoyed mentoring them and explaining what the differences were from flying in the simulators versus actually being in space. The rookies were responsible for designing our crew patch, and it included many of the payload representations on it. The payload commander duties were huge on this mission, as we had over twenty payloads, in addition to the MFD Japanese robotic arm and the CRISTA-SPAS. So, it was a busy patch!

NASA was preparing to build the ISS in 1998, so there were many studies on what kind of robotics were needed for assembly of the ISS. This delighted me, as Steve and I would be performing additional robotics with the CANADARM RMS, in addition to deploying and retrieving the CRISTA-SPAS and operating the Japanese MFD.

Our Space Shuttle simulations in the SMS began in early 1997, and we had a long training syllabus to perform the rendezvous with the CRISTA-SPAS. Steve and I spent many hours training on the CANADARM RMS. Higher management at NASA decided that I should deploy and retrieve the CRISTA-SPAS, as they wanted an experienced arm operator to perform these critical duties. Steve would be operating the CANADARM RMS a lot during the various maneuvers that we were going to do in support of ISS assembly, and we would share "flying" duties on the MFD Japanese robotic arm as well.

Our launch was scheduled for August 1997, so we had a busy training schedule. It was very enjoyable, as the chemistry among the crew was fun and supportive. Unfortunately, Jeff Ashby asked to be removed from the crew so he could care for a terminally ill family member. Therefore, in March 1997, only four months before our scheduled launch, Jeff was replaced by Kent "Rommel" Rominger. This was hard on all of us but especially Jeff.

Rommel was a very nice guy and a super addition to the crew as he fit right in. He had recently flown on a Space Shuttle mission in November 1996 with a WSF deployment and rendezvous, so he was familiar with rendezvous and proximity operations procedures. Our flight would be his third journey into space.

As a crew, we continued our intense training, relieving the stress with humor and fun. I was lucky that, once again, the mission was a high inclination orbit so that we could fly over Japan. This meant that we would be flying over most of the Earth, from north to south, and this time, I would not be stuck in a module! We spent a lot of time studying what Earth observations we would make, and I was looking forward to more time at the window.

Another one of my assigned duties was photo and television responsibility, which made sense because many of the payloads had extensive photography requirements. On this flight, we carried a bevy of cameras—the Linhof camera, three Nikon cameras, three Hasselblad cameras, and four camcorders. Whew!

Steve and I had one more trip to Japan to train on the MFD hardware simulator, and we went with Tinch-san in April 1997 during cherry blossom season. While there, I was able to reconnect with some friends I had made on my first mission as well as one of my quilting friends, Sunoko. Steve and I had a wonderful trip, and we felt like we were prepared to fly MFD on orbit. Our Japanese training team led by Sato-san as well as Tinch-san and Sho felt we were ready to fly. I introduced Steve to Japanese curry rice, and we purchased some rice cakes and packaged Japanese curry to eat during our spaceflight.

As the launch date grew near, we did the things to which I had become accustomed in getting ready for a launch—making the list of special guests and family who would attend the launch, choosing a menu, and selecting what items would go in my PFK and OFK. We were also allowed to carry two personal items, such as a school pennant or a favorite bumper sticker, to display in the crew cabin, and twelve personal compact discs of music.

The only items we were able to keep as our own were the crew shirts. We picked out a couple of shirts that we all had in common so we would have a "uniform" for things like rendezvous day. And we were the first crew to have matching boxer shorts with our crew patch! By my third mission, I knew what clothing worked best for me; I generally chose short sleeve and long sleeve cotton shirts and a cobblecloth cotton pullover sweater. We could have our shirts and sweaters monogrammed with our name, our initials, our mission number, or our patch. Our mission was scheduled for eleven days, so we were allowed to choose twelve different shirts.

Our crew also went through the Space Shuttle escape training and firefighting training as well as practiced driving the armored truck at KSC, rappelling over the side of the Shuttle mockup at JSC, and deploying the pole out of the middeck of the orbiter mockup. We also simulated bailing out of the Shuttle on this pole in the event of an emergency, among other things.

By my third mission, a new huge water tank had been built at JSC for EVA training. It was large enough to hold two complete Space Shuttle payload bay mockups and was needed to start training for the assembly of the ISS. Called the Neutral Buoyancy Laboratory (NBL), it was used to practice our dinghy and parachute training while we were in our ACESs. All in all, I was much more comfortable with these activities and wanted to make sure that the rookies were adequately trained and understood what they needed to know.

Beamer and Steve would be the EVA crewmembers for this flight. I was the intravehicular activity crewmember who would help them suit up if there was an EVA needed. Therefore, I was in the control room of the NBL for all their practice runs. I thoroughly enjoyed this.

There was some travel needed for experiment training but not as much as we'd had during my first two missions. Most of our training was conducted at JSC or at KSC. However, Steve and I traveled to White Sands, New Mexico, to observe the Hale-Bopp comet in the wee (cold) hours of the morning. On our flight, we had an ultraviolet telescope that was going to take pictures of the comet through the middeck crew hatch window. We still flew a lot in the T-38s as needed for proficiency and for going to places like KSC for training, Goddard Space Flight Center for our secondary payloads training for International Extreme Ultraviolet Hitchhiker (IEH) and Technology Applications and Science, Huntsville for training at UAH, and El Paso for Curt and Rommel to fly the Shuttle Training Aircraft practice approaches.

During training, Beamer and I really hit it off and became good friends. We worked well as "back-seaters" in the Shuttle simulators, and we were partners in egressing the Shuttle if there was an emergency. We both love chocolate, so if either of us was around chocolate, we would bring a sample for the other one. We each stowed one big bar of chocolate for each day of our spaceflight (Figure 21-2)!

I was honored and humbled to have the widow of Wernher von Braun and his family come to the launch. I'd invited them to my two prior launches, but they weren't able to come until my third launch. Dr. von Braun's three children and their families, along with Mrs. von Braun, were my special VIP guests. Mrs. von Braun wanted to meet as many astronauts as possible at the Banana Creek viewing site, and my parents were happy to take them around. The astronauts were thrilled to meet Mrs. von Braun! I had known the von Brauns ever since I'd gone to Randolph School with Margrit von Braun, and it was meaningful for them to come to my launch.

Figure 21-2: Bob Curbeam (Beamer) and I enjoying some chocolate on the flight deck of Discovery, STS-85 (NASA photo)

On August 7, 1997, the launch countdown went smoothly, and I was a good luck charm once again as we launched on time—and launched two big candles for Rommel's birthday—at 10:41 a.m. During the countdown, all I could think about was the deployment of the CRISTA-SPAS as soon as the post-insertion checklist was over. That was a big responsibility, but first, I had to climb out of my seat and take pictures of the ET once MECO occurred.

I was comfortable as one could be in the bulky ACES due to my familiarity with the launch sequence and knowing what to expect as well as the constant cooling in the suit. Therefore, I enjoyed the launch so much more and was not apprehensive about it. It was enjoyable to see the three rookies' expressions and hear their squeals of joy once we were floating out of our seats after MECO. Although we did not have a module on this flight, we had an exterior airlock outside of the middeck area. This gave us more space to store our seats and our ACESs and gave me a nice changing room away from the guys. Actually, I felt like I was camping out with five big brothers, and it was no big deal.

Of course, one of the first things we had to do was set up the potty. I haven't mentioned it before, but in one corner of the middeck, there was a toilet called the waste collection system. Basically, there was a hose with suction, and that is what we used for urination. If we had to do number two, we sat on the toilet (making sure that the bars were across our legs so that we would not float away!), opened a door, and did our business. There was a little bit of suction to take care of what gravity would normally take care of and a spinning drum that we affectionately called "the slinger." The slinger would then be opened to the vacuum of space, and we had freeze-dried crystals of our excrement. The urine was also freeze-dried and dumped overboard, creating the most beautiful crystals ever! Any waste such as toilet paper would be sealed up in a bag and stowed. After that was taken care of, setting up the galley, the cabin air fan, and other housekeeping duties kept the post-insertion timeframe a very busy one.

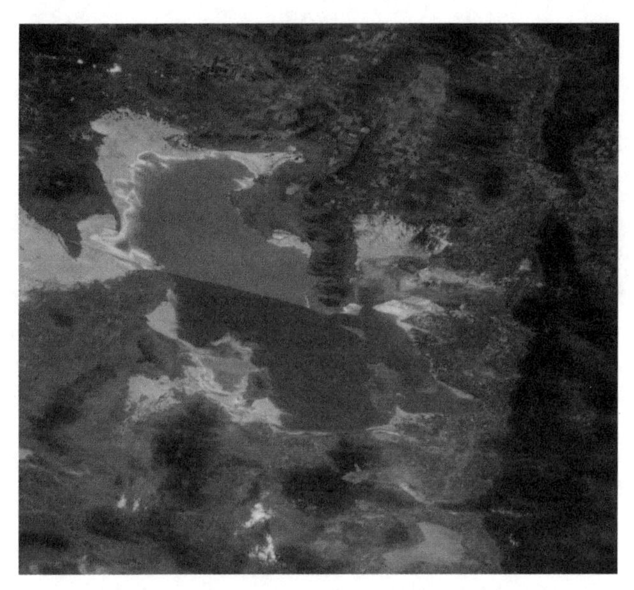

Figure 21-3: Great Salt Lake, Utah, as photographed from space by STS-85 crew (NASA photo)

As soon as we were finished with our post-insertion checklist and had opened the payload bay doors, we were on orbit where we could see the Earth go by. One of the first things I saw out the window was the United States. There was a huge lake that was divided into two halves—one half was deep blue, and the other half was bright red (Figure 21-3). What could that be? I found out later that it was the Great Salt Lake in Utah, and it was divided into a shallow half and a deeper half by the Lucin Cutoff, a road/railway built in 1959. The northern half was a deep red because, at that time of year, a type of algae flourished with the high salinity. The dissolved minerals, turbidity, and microorganisms that can survive in saline water give the lake its varying colors. I had never heard about that before or seen it, and it was spectacular. Who knew?

Figure 21-4: Payload bay of Discovery STS-85 with Japanese Manipulator Flight Demonstration arm in the foreground and the CRISTA-SPAS about to be deployed from the CANADARM RMS (NASA photo)

I was feeling great with no SAS and was busy getting ready to deploy the CRISTA-SPAS. As I began power-up of the CANADARM RMS, uncradling it and "flying" it around with Steve by my side, the maneuvers felt very familiar. The end effector of the RMS had wire snares, and with a flip of a switch, they grabbed the grapple fixture that was on the CRISTA-SPAS. The CANADARM RMS was rock steady, and I was able to unberth the CRISTA-SPAS to ready it for deployment. This took a lot of coordination with the orbiter crew (pilot, commander, and MS2) and the ground, as the orbiter had to be in the right orientation and altitude for deployment. A picture of our payload bay prior to deployment is shown in Figure 21-4, which depicts the MFD in the foreground and the CRISTA-SPAS on the end of the arm. The deployment of the CRISTA-SPAS went off without a hitch only six hours after our Space Shuttle launch. I was thrilled and relieved to successfully send it on its way.

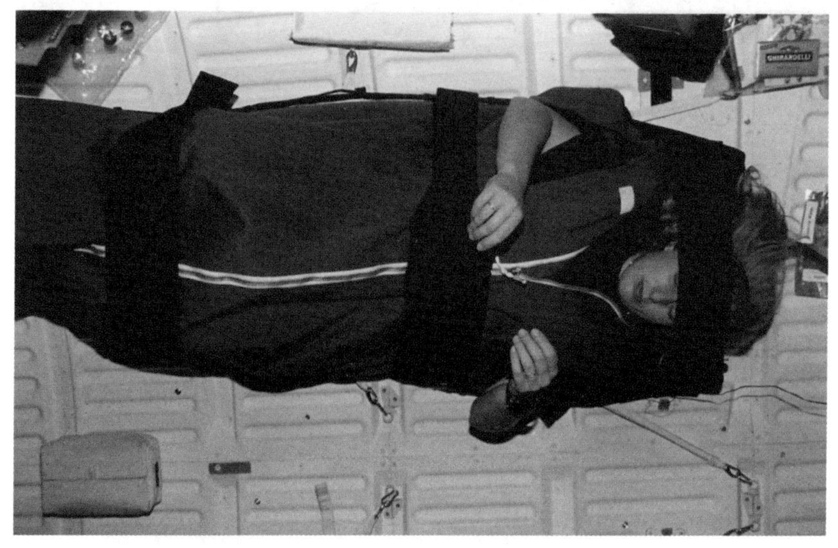

Figure 21-5: Jan sleeping in her sleeping bag on the middeck of Discovery, STS-85 (NASA photo)

After CRISTA-SPAS was deployed, the crew set up our home away from home for the next eleven days. We had to configure the payloads as well as set up thirteen laptop computers for the payloads. Plus, we made sure that the rookies had a chance to look in awe at our beautiful Earth.

About ten hours after launch, we began preparations for our single-shift mission by having dinner and climbing into our sleeping bags. We hung them wherever we could. Some of the crew slept on the flight deck, and some slept on the middeck. I chose to sleep on the middeck and strapped my head to the pillow with a big piece of Velcro! I also wore a mask over my eyes and never had trouble sleeping while I was floating in space (Figure 21-5).

The next few days were busy with experiments and robotics activities. I was pleasantly surprised at how well the MFD operated with only a few glitches. Steve was helping me and operating the MFD as we took turns flying it. Figure 21-6 shows a picture of me operating the MFD with my MS1 Japanese headband!

Figure 21-6: Jan operating the Japanese Manipulator Flight Demonstration (MFD) arm on the aft flight deck of Discovery, STS-85 (NASA photo)

Over the course of several days, Steve and I spent a total of about twenty hours supporting the MFD payload, including intricately opening a door with the arm and moving boxes around the payload bay with it. There was also some time on flight day seven when the Japanese robotic arm was controlled by the Japanese on the ground. It was a successful payload, and I am so glad to have played a part in the testing of the arm that is now the Small Fine Arm on the ISS. It is attached to the longer Japanese arm, and the duo is used on the ISS to take experiments off the Japanese Experiment Module pallet and bring them inside the airlock and the ISS for experimentation.

One of the interesting experiments in the middeck was the Solid Surface Combustion Experiment. The PI for this was Bob Altenkirch, who would later become the president of the UAH. Believe it or not, Rommel and Curt started a fire (deliberately) in a furnace! With a normal flame experiencing Earth's gravity, the teardrop shape is formed by some of the gases around the center of the flame being heavier than others. I think it is fascinating that, without gravity, the flame forms a sphere instead of a teardrop. Hopefully by doing experiments like this, we can learn more about fires and flame propagation.

We did not have as many media events or public affairs interviews during this mission compared to my first two. However, due to the workload of so many experiments, our days were full, and we were exhausted. There were quite a few maneuvers of the orbiter needed to observe the sun, Jupiter, stars, and other heavenly bodies, and this kept Curt and Rommel busy. It was so much fun flying with the three rookies because they enthusiastically took on any task and took care of most of the housekeeping duties needed to keep things in shipshape.

Flight day ten was the big day, as it was rendezvous day with the CRISTA-SPAS, so we put on our matching red and white striped rugby shirts. Curt and Rommel could fly the orbiter from the aft flight deck behind where the pilot and commander normally sat. Beamer sat in the front and entered commands onto the keypads that drove our computer displays.

Not only did Curt, Rommel, and Beamer gracefully conduct several maneuvers to bring the orbiter in the vicinity of the CRISTA-SPAS, but I was put to the test of grappling the satellite. I had butterflies in my stomach as we approached the CRISTA-SPAS. I had maneuvered the CANDARM RMS to the right position so that I could grapple the CRISTA-SPAS and was intently looking at the television on the end of the arm to determine where the grapple fixture was. I saw it! At that point, Curt and Rommel had stabilized the orbiter in free drift, and I slowly and gradually moved the arm toward the CRISTA-SPAS. I had to be careful not to inadvertently tip the satellite and send it hurling toward the orbiter or allow it to be lost in outer space. I also had to watch the angles of the arm so that I did not hit any reach limits of each joint or hit a singularity, which would cause the arm to automatically reconfigure.

It was a successful grapple, and I calmly relayed to Houston mission control, "Houston, Discovery. We show a good grapple," even though my heart was still racing! Everyone on the crew hooted and hollered, slapped me on the back, and gave me high fives. It was especially gratifying, and I was so relieved. A picture taken of us in our rendezvous shirts and our matching boxer shorts for a celebration is shown in Figure 21-7.

Figure 21-7: STS-85 crew in the Discovery middeck after a successful rendezvous and capture of the CRISTA-SPAS scientific satellite (NASA photo)

The next day was supposed to be our last, and I was pretty bummed out that I had not yet seen my hometown on any of my three spaceflights. On STS-85, everyone had seen their hometowns, and I was determined to get a photograph during our last pass over Huntsville. The time there was around 9 a.m., and each person on the crew was at the window with a different camera. Traveling at five miles a second, we had to be quick in recognizing the geography. I excitedly recognized Jackson, Mississippi, and then Birmingham, Alabama, so I gave the crew a heads up that we were close to Huntsville. There it was!

One of the most exciting times in space for me was seeing my hometown from space. It was beautiful with a rolling fog still present in the foothills around the city and the unmistakable airport runways. Therefore, we have lots of pictures from that momentous occasion, and I felt full inside.

We were busy stowing experiments, equipment, cameras, etc., when we found out that our spaceflight had been extended a day due to bad weather at our KSC landing site. This was a great luxury because after a successful and busy mission, we finally had time to relax and look out the window a little bit. We had already taken our official crew picture but had more time to take pictures of the Earth. We were happy to be in space another day!

I am usually asked by people if I have any funny stories about what happened in space. Most of my total of twenty-eight days in space were very busy without a lot of time for frivolity. Every minute of our time was scheduled, and unlike the astronauts on board the ISS, we only had a week or two on-orbit instead of months. However, the hardest I ever laughed while I was in space was during an interview of our crew with some school children during our last full day on orbit.

All of us were tired and probably slap-happy, but we answered the school children's questions with aplomb and energy. However, one child asked Curt, "Does your hair grow faster in space?" Good question! Some of the guys on our crew had, shall we say, receding hairlines or thinning spots. When this question was asked, we all burst out laughing with good belly laughs. Curt could not contain himself and was laughing so hard he was crying. I am sure that everyone who was listening was wondering what was going on, as the response was silent. Finally, when we all pulled ourselves together, Curt answered with something like, "I don't know. I'll let you know when I find out!"

Finally, after waking up on flight day twelve, it was time to put on our pumpkin suits and come home. The rookies were surprised at how painful and difficult it was to put them on while on orbit, but we were able to get the payload bay doors closed and get everything stowed before we put on the suits. Then, we dutifully drank our salt water, and I advised the rookies on what had helped me avoid becoming lightheaded after landing.

After our deorbit burn, it was a typical entry with a bit of a rocky road and a few g's experienced, but once Curt or Rommel was flying the vehicle, it was a smooth ride. For the third time, I was in a Space Shuttle landing where we were supposed to land in Florida (Figure 21-8). Again, I was the lucky charm! I could not believe all my missions had launched in the first minute of the first day of the launch window, and all of them had landed at KSC. Whenever I climbed into the Space Shuttle, it took off! And whenever I climbed out of the Space Shuttle, I was at KSC.

Figure 21-8: STS-85 landing at Kennedy Space Center, August 19, 1997 (NASA photo)

My mother was always supportive of my spaceflight missions and was there at every launch and landing as well as each step along the way. Fortunately, I'd been able to take her onto the launch pad at KSC before the launch so she could see the Space Shuttle up close. I know it was hard for her to see her firstborn go into the risky environment of space travel, just like it had been difficult for her to see her husband go off to a dangerous war. She was such a positive person, and she knew that we were both following a dream and serving our country in a special way, so she supported us. Once this mission was over, she welcomed me home one more time, feeling relieved and happy. I can't imagine her joy when my father returned home after being a pilot in combat and being a POW, but I am guessing that the emotions were similar to when I completed my spaceflights, and she welcomed me back with loving, open arms.

Our crew was so happy that all aspects of our mission had been successful, and we were elated to walk around our beautiful Discovery after Curt landed our craft so well on the SLF. We were then able to see our friends and family who'd come to watch the landing in Florida, after which we made our way back to Houston in the Gulfstream aircraft. A crowd welcomed us home at Ellington Field, and it was good to be back on planet Earth.

My return was bittersweet, as I did not expect to fly again—by choice. I was fortunate to have been part of three spaceflight missions that were international in nature. My third flight was no exception, as we had post-flight trips to Germany, Italy, Japan, and Canada.

In late October, we went on our first international post-flight trip to Germany, as the CRISTA-SPAS satellite was sponsored by the German space agency, and they were thrilled that we'd been able to deploy and retrieve the satellite successfully. Although they were still poring over the data, they knew it was a successful mission and were very happy with our crew.

Our hosts/hostesses were gracious and provided our crew with a Mercedes and BMW so we could drive ourselves to Italy from Germany on the Autobahn. That was probably riskier than being on the Space Shuttle, but we made it to Italy with no problem. One of our cars took a detour to the Italian Alps, and they skied above the clouds before arriving in Trieste.

The Italians were responsible for some of the experiments on the IEH truss and provided us with a warm welcome, complete with extraordinary wine and food. The mayor of Trieste at the time was Riccardo Illy, of Illy coffee fame, so we also enjoyed coffee, along with other delicacies. We had an amazing trip to Europe, and I am thankful for the partnerships that NASA has with the Italian Space Agency and the ESA.

In mid-November, our crew went to Japan to celebrate and debrief the successful mission of the MFD. The Japanese were very happy with the arm's performance and the crew operation. It was heartwarming to be back in Japan and to celebrate with a new set of Japanese colleagues.

The Japanese have been great partners in building the ISS and providing top-notch astronauts to the program.

After the new year rang in 1998, our crew made one more international trip to Canada, the home of crewmember Bjarni. He'd been born in Iceland and, at the age of eight years old, he and his family had moved to Vancouver, British Columbia. We had a remarkable visit to Vancouver, followed up with a trip to Whistler and Calgary to go snow skiing. We made wonderful memories and had such a good time.

I feel blessed to have been on three international spaceflight missions that were not only interesting and fulfilling from a personal standpoint but also historic in nature. They were successful missions and had fantastic crews. A summary of these three missions is in Table IV of the appendix.

The missions served our country's space program in a meaningful way that furthered our technology and expanded the boundaries of science in space. Personally, I'd fulfilled my dream of flying into space, and it combined all the things I loved—space, science, engineering, people, mental challenge, international relations, physical activity, exploration, and service to our country. What more could I have asked for?

It was humbling to be one of the lucky few who traveled into space, as thousands of people around the world gave their heart and expertise to make it possible for us to have safe and successful missions. I am thankful for their hard work, and I hope they know how much the flight crews appreciate everything they have done and continue to do for our country's space program and space programs around the world.

At this point, I was ready for my next challenge and eager to achieve my next goal—to become an executive at NASA. On to the next chapter!

Chapter Twenty-Two

There Is Life After Being an Astronaut

Shortly after the debriefs, post-flight trips, and speaking engagements, I was made aware of a NASA executive position in the SES at NASA headquarters in Washington, D.C. I would be reporting to Fred Gregory, who was chief of the Office of Safety and Mission Assurance (SMA). I interviewed for the job, and after a lengthy screening process to get into the SES, I was selected! Only a few astronauts had made it into the SES, and I'd worked hard and done some extra activities to qualify.

The job for which I was selected in early 1998 was the Human Exploration and Development of Space (HEDS) director of independent assurance. My group and I were responsible for making sure that the HEDS programs, Space Shuttle and ISS, had independent analyses and testing to make sure they were safe to fly humans into space. The job was challenging but very satisfying, as I believe we were able to make constructive efforts to bolster what extra analyses, engineering, and testing the ISS program was not able to do. I enjoyed managing the people, working with the program executives, and being an essential part of the program as we were getting ready to assemble the ISS. I was on programmatic boards, was traveling to Russia (with whom we were partnering to build the ISS), and was in critical meetings necessary to overcome technical challenges.

The first launch for the assembly of the ISS was December 1998, and the Space Shuttle STS-88 mission carried the first U.S. element, the Unity node, which docked with Russian ISS elements already in space. It was an exciting time for our space program, and I was very happy to be in the middle of helping make it happen. I felt like it was a natural progression for me as an astronaut to be involved in a different way with an international mission.

In late 1998, a new center director for MSFC in Huntsville, Alabama, was selected—Art Stephenson. Art approached me in 1999 about coming to MSFC as part of his new organization. MSFC was where I'd started my career, and my parents still lived in Huntsville. As they were in their mid-seventies, I was anxious to be with them as they grew older. We would be able to do things together, and I would be able to help them when needed. Art and I discussed the reorganization that he'd made at MSFC; then, he made a few tweaks after our discussion and put the development of all the ISS hardware and science organizations into one organization. From there, he offered me a position as the deputy of the Flight Projects Directorate where Axel Roth was the director.

I said my goodbyes in Houston, had lots of parties and tears, and drove to Huntsville. In July of 1999, I started my new job at MSFC and couldn't have been happier. The Chandra X-ray program was managed in our directorate, so I went to the Space Shuttle launch of the Chandra, where the first woman Shuttle commander, Eileen Collins, was heading the mission. It was also the thirtieth anniversary of the first moon landing that month, and celebrations at MSFC ensued. I felt right at home back at MSFC.

Axel and I set about organizing the Flight Projects Directorate, and it was during a busy time for ISS. Our directorate was responsible for setting up a mission control type facility to manage the science for the ISS. This involved training the crew, writing the procedures, working with the scientists, building the ground systems, training the flight controllers, planning the day-to-day science operations for the ISS, etc. The facility and the training needed to be finished before we had a crew on the ISS when the U.S. Laboratory Module was installed. We had a great group of people working on all of this, and I was confident that we would meet the deadline.

Our organization was also responsible for building several pieces of hardware that would be installed onto the ISS, which would be an orbiting laboratory with continuous occupation for the crew. Therefore, many scientific experiments would be conducted, and our hardware would enable this breakthrough research. We built the EXPRESS science racks, which would contain the experiments, the glovebox for experiments that needed a contained enclosure, and several of the pallets that would be used to carry hardware on the Space Shuttle to the ISS. It was the perfect job for me, as I had experience with flight hardware and on-orbit flight operations and had good relations with JSC, KSC, and our international partner countries.I was also able to convince ISS Deputy Program Manager Jay Greene to have monthly ISS meetings at MSFC, and our ISS colleagues at KSC and JSC also came to the meetings. These meetings were a huge step in building the inter-center relationships necessary to make the ISS program a success.

In January 2001, Axel was moved up to be associate director of MSFC, and I was named director of the Flight Projects Directorate. Art was a true leader who took time to teach leadership to his SES employees, and being one of his direct reports made me appreciate him even more. I was glad to be a member of his senior staff and learn more about the operation of MSFC.

As we were constructing the ISS piece by piece in space, our organization was launching hardware on almost every Space Shuttle launch. I participated in the flight certifications of the hardware and was thrilled to be a part of the flight readiness reviews. I sat with my KSC colleague, Tip Talone, and others in the firing room at the KSC Launch Control Complex for each one of those launches.

In February 2001, the U.S. Laboratory Module arrived at the ISS, and we had a grand opening of the Payload Operations Center at MSFC. We were ready for twenty-four hours a day, seven days a week operation of the control center for the science on the ISS. Starting March 2001, the ISS has had a crew on board, and the Payload Operations Center has been coordinating the science on it ever since March 2001.

Everything was going well with the construction of the ISS, and a rare Space Shuttle mission STS-107 was scheduled that would be a pure science and technology mission in a Spacehab double module on the Columbia Space Shuttle flight. Our Flight Projects Directorate had an experiment on STS-107 that involved testing a distillation apparatus for our regenerative ECLS system.

When I was in Houston during the training of the crew on STS-107, I had breakfast with one of my dear friends on the crew, Kalpana Chawla, whom we called KC. She always had a big smile, but she had an especially big smile that morning as she told me how STS-107 was the perfect mission for her because she really enjoyed the science on the mission and was learning so much during her training. I watched with glee as KC and others on her crew not only launched and successfully carried out a multitude of international experiments but also enjoyed floating in space and having fun.

On February 1, 2003, as Columbia was on her way home, she broke up over Texas, killing the entire crew. It was a Saturday, and Art called in his senior staff for a time of prayer and to talk with others in NASA about next steps. Critical personnel were deployed to Texas to start the unpleasant job of recovering Columbia and her crew. I knew every member of the crew but was especially close to KC. I had also flown T-38s with Commander Rick Husband. I was in shock and mourned their loss along with the rest of our nation. It was a heartbreaking loss personally and professionally.

I was able to go to some of the Columbia recovery sites in Texas and gave talks and signed autographs for the hundreds of people who were involved in the recovery. Most of them were firemen from around the country and Puerto Rico who were coordinated by personnel from the U.S. Department of Agriculture (USDA). The USDA was prepared to respond to disasters and deployed ready-made communities out in the middle of nowhere when there were forest fires.

I continued to work on ISS projects in the Flight Projects Directorate. However, the Space Shuttle flights were suspended after the accident, so we could not take more hardware to the ISS. Nonetheless, we continued science operations on the ISS and continued to successfully operate the Payload Operations Center.

Meanwhile, NASA was in turmoil with an uncertain future. The president had assigned a board, the Columbia Accident Investigation Board (CAIB), to investigate the accident. It was planning on releasing its report in August 2003.

About a week before the CAIB report was scheduled to be released, the head of the MSFC Office of SMA resigned. It was a critical position at that time for the center, and new Center Director David King asked if I would step in as the director of SMA. I did not hesitate to say "yes," even though I was perfectly happy to keep working on the ISS as the director of a much larger organization. However, I felt it was my duty to help MSFC and NASA by being a part of getting the Space Shuttle flying again.

It had been determined that a chunk of foam insulation on the ET of Columbia had come loose during the launch and had hit the wing of the orbiter. The resulting gaping hole had allowed hot gases to enter the wing during Shuttle reentry, causing the failure of the wing and the destruction of the vehicle. The ET was a MSFC project, and it was scrutinized during the return to flight activities. MSFC SMA was at the center of attention to make sure that it was fixed.

The next two years were packed with significant meetings, rebuilding of the SMA organization, redesign of the ET and its foam processing, and personal psychological pressure for me. By the time we launched the Space Shuttle successfully in July 2005, I was burned out and ready to do something else, so I resigned from my position as director of SMA at MSFC and retired from NASA in October 2005.

After my retirement, my life was shaken by the illnesses of my mother and stepfather. My mother had complications due to medication for her arthritis, and she died in January 2006. She was my best friend, my confidante, and my chief cheerleader. She had made such a positive difference in my life. I miss her.

My stepfather moved back to his hometown of Fort Worth in Texas soon after my mother died. He later experienced cancer and died in 2009. I am grateful for him raising me with my mother ever since I was five years old.

Also in 2006, I started dating an old friend from my Randolph School days, Dick Richardson. We were set up on a blind date that wasn't very blind, and he was a true soulmate for me. On the day after Christmas 2006, we married, and he became my lifelong partner. Along with him came two bonus children, Frances and Patrick, and they and their families have richly blessed my life.

Shortly before my retirement from NASA, I was approached by Jacobs, which provided engineering services for MSFC. I agreed to go to work for them, as I was familiar with the company that had been one of my contractors when I was the director of the Flight Projects Directorate. As deputy general manager for the Jacobs contract, I would be involved in managing over one thousand engineers and technicians who provided in-house engineering support to MSFC.

NASA was building a new rocket named Space Launch System (SLS), which was being designed and managed out of MSFC. As one of its major contractors, we were heavily involved in the in-house development of hardware and software for SLS. The new rocket, which would use the remaining SSMEs and the SRB segments from the Space Shuttle program, required a new fuel tank to be designed and built, and our contract was heavily involved in this design of the core stage. JSC would be responsible for the crew capsule atop the SLS rocket, and it would be called Orion.

In 2017, I was approached by Bastion Technologies at MSFC to be the program manager on the SMA contract. When I was director of SMA, I was very familiar with my contractor, which was later bought out by a family member and called Bastion Technologies. Therefore, I knew about the history and integrity of the small business and was happy to join their family. I enjoyed being back in SMA and had a terrific team and a superb customer with which to work. About 75 percent of our workforce was working on the SLS program from a safety and quality assurance perspective. It was a good move for me, and I retired from this position in January 2020.

I am thankful beyond measure for the opportunities I have had in my education and career. I don't think the divorce of my parents caused any long-term negative effects on me but motivated me to achieve my goals. I have worked hard, but I have had (mostly) good bosses, good organizations, wonderful people, and great projects over the years. I hope to "pay it forward" to the next generation of explorers and give them the opportunities and inspiration that I had.

After my retirement, the COVID pandemic hit the entire world, and I was left with canceled or deferred plans and not much to do but quilt. In April of 2020, my sister Darby Smotherman also had extra time on her hands and was able to scan my father's POW Wartime Log and send it to me. Suddenly, I had a project! This project will continue when I publish the Wartime Log and the "PENNY" comic strips in a companion volume to this book.

I spent three years doing research, traveling to libraries, going to East Anglia, visiting Germany, and writing. It has been a time of discovery for me—not just discovering my father's history but also discovering myself. For the first time in my life, I feel complete, knowing my roots and being able to share with others what I could not share for so many years.

With this information I learned about my father, I was able to have the POW Medal awarded to him posthumously by Alabama Senator Tommy Tuberville. Well done, Colonel Benjamin Franklin Smotherman, well done. Thank you, Daddy Ben, for your service from the bottom of our hearts and from a grateful nation.

Acknowledgments

This labor of love, which took three years, would not have happened without the support and wise counsel of many. There are many people who have shared their knowledge and experiences with me, and I appreciate their encouragement in the writing of this book. Although there are too many people to mention here, these stand out.

For the research I conducted, I would like to thank the expert help from the researchers at the Air Force Historical Research Agency at Maxwell Air Force Base (AFB), who found microfiche records of the 92nd BG, 327th BS, as well as lengthy reports from my father's seven combat missions. Reproduced onto compact discs for me, this information was invaluable. Judy Roddy, archivist at the Mighty 8th Air Force Research Library in Pooler, Georgia, was also helpful in uncovering great details about the YB-40, my father, and his time in Stalag Luft III. At the Air Force Academy McDermott Library Albert P. Clark POW Collection in Colorado Springs, Colorado, special collections archival technician Ruth Kindreich spent hours pulling files, scanning images, and providing resources for me. Bub Clark's personal notes and scrapbook retold the story of him being imprisoned along with my father, and I am indebted to Ruth for helping me uncover these gems. Lieutenant General (retired) Susan Helms and Colonel (retired) Gary Payton helped me to gain access to the Air Force Academy, and I would not have been able to do this valuable research without their assistance.

The National WWII Museum in New Orleans, Louisiana, is a tremendous asset to the WWII community, and I took advantage of their POW course, led by senior curator Kim Guise. She not only conducted an excellent course but also answered questions and pointed me in the direction of other resources available. In May 2022, my husband and I went on the "Masters of the Air" tour of East Anglia, led by the National WWII Museum's head of educational travel, Nathan Huegen. This tour opened my eyes to life in England for the 8th Air Force, and Nathan made a special effort to take me to my father's base in Alconbury. The tours of air bases, museums, and pubs made the experience that my father had come alive.

I am also deeply grateful to Gunther Strehle in Moosburg, Germany, for the very informative and wonderful tour of the remaining buildings of Stalag VIIA and to Peter den Tek of the Sky of Hope Museum in Holland for the photograph of my father's crashed plane as well as other articles in the Dutch press about my father's crash. Sarah Batemen Testa worked tirelessly to produce the maps of my father's missions and his journey from Stalag Luft III to Moosburg. She accurately portrayed towns and borders from the era and lent a youthful enthusiasm to this project. Thank you, dear Sarah.

One of the highlights in preparing to write this book was a ride in the Commemorative Air Force's (CAF's) B-17G Texas Raiders. I will never forget the pilot, Len Root, making a special effort to let me sit in the pilot's seat and exercise the flight controls (on the ground) so that I could experience some of the sensations that my father felt while piloting this massive bird. The CAF historian, Kevin Michels, was enthralled with my father's story and answered many of my questions about the life and times of B-17s and their crews. Len and Kevin sadly passed away in their beloved Texas Raiders during the writing of this book.

I am fortunate to have immensely talented friends and colleagues who have lent their support, advice, and review of this book. In particular, authors such as Homer Hickam, a friend and NASA colleague; astronaut Tom Jones; new friend Ross Greene; and cousin Martha Burns provided much-needed guidance in the journey of publishing and were tireless in answering my many questions. Along with us on our tour of England, *Masters of the Air* author Don Miller was very helpful in not only answering questions but also providing encouragement and support during the writing process. I especially appreciate the expertise and editing of my manuscript by Marilyn Jeffers Walton, who is the daughter of a POW and an expert about Stalag Luft III, among many other subjects. Marilyn introduced me to Ed Reniere from Belgium, who did extensive research to find information about my father's crew and colleagues. I appreciate this so much and look forward to contacting the families of these special friends.

Heartfelt thanks also go out to Patti Schoborg, who facilitated my meeting with her father, navigator Bob Doolan, in Cincinnati. Bob was the only person I ever met who knew my father as an aviator, flew as a crew-member on the YB-40, was in his same BG and BS in England, and was a POW in Stalag Luft III. It was a delight to meet Patti and her father, as she and I have a special connection as the daughters of very courageous men.

In the course of writing this book, I have been supported in innumerable ways by my editor, Lauren Green, whose patience and assistance helped this engineer turn words into a worthwhile book. Thanks also to Kayleigh Rucinski and the team at Ballast Books for guiding me through the process of putting this book together. My agent, Burke Allen, along with Jeff Johnson from Allen Media Strategies, were indispensable in being my "sherpas" through the publication process for this first-time author.

I have been blessed with incredible parents—my mother, father, and stepfather who have helped to make me the person I am today. I dedicated this book to my family, who have been inordinately supportive and patient, including my husband, Dick Richardson; brother- and sister-in-law, Jim and Nancy Richardson; brother, Ron Davis, and his wife, Gina; sister, Darby Smotherman; niece, Abby Davis; and children, Frances McCarty and Patrick Richardson and their families. A book dedication is a small token of my gratitude but hopefully signifies my love and appreciation of them.

References

*** Primary References**

1. *92nd Bombardment Group reports. Microfiche, US Air Force Historical Research Agency, 1943.
2. *327th Bomb Squadron reports. Microfiche, US Air Force Historical Research Agency, 1943.
3. *American Prisoners of War in Germany*. Military Intelligence Service, War Department, 15 July 1944 and 1 November 1945.
4. Fort Worth Texas City Directory, page 755. 1936.
5. *Fuwatto '91 First Material Processing Test*, NASDA National Space Development Agency of Japan, 1991.
6. "Hospitalization and Evacuation of Recovered Allied Military Personnel." Administrative Memorandum Number 48. WW2 US Medical Research Centre Supreme Headquarters Allied Expeditionary Force, APO 757, February 26, 1945. (Source: med-dept.com website)
7. *Space Shuttle Mission STS-47 Press Kit*. National Aeronautics and Space Administration, September 1992.
8. *Space Shuttle Mission STS-60 Press Kit.* National Aeronautics and Space Administration, February 1994.
9. *Space Shuttle Mission STS-85 Press Kit*. National Aeronautics and Space Administration, August 1997.
10. *Spacelab J, Microgravity and Life Sciences*. National Aeronautics and Space Administration, 1992.
11. "Stalag Luft III, The Secret War." *The Longest Mission*. The Association of Former Prisoners of Stalag Luft III, 1995.
12. "Survivor Depicts Andrews Crash." *The New York Times*. May 7, 1943.

13. "Yanks Leave Trondheim in Flames". *Dallas Morning News*. Sunday, July 25, 1943, London Associated Press.
14. Ashcroft, Dr. Bruce, Staff Historian. *We Wanted Wings: A History of the Aviation Cadet Program*. HQ AETC Office of History and Research, 2005.
15. Birdsall, Steve. *Pride of Seattle – The Story of the First 300 B-17Fs*. Squadron/Signal Publications, Inc.,1998.
16. Bolden, Charles F. *Edited Oral History Transcript.* Interviewed by Sandra Johnson. NASA Johnson Space Center Oral History Project, Houston Texas, January 2004.
17. Calvarese, Pasquale (Patsy) Louis. "Journal". 96th BG, 413 Squadron (Source: Albert P. Clark Collection of the US Air Force Academy McDermott Library)
18. *Clark, Albert P. *33 Months as a POW in Stalag Luft III, A World War II Airman Tells His Story*. Fulcrum Publishing, Golden Colorado, 2004.
19. *Clark, Albert P. *Scrapbook*. Friends of the AFA Library, (Source: Albert P. Clark Collection of the US Air Force Academy McDermott Library).
20. *Crosby, Harry H. *A Wing and a Prayer*. HarperPrism, 1994.
21. Cross, Robin. *The Bombers: The Illustrated Story of Offensive Strategy and Tactics in the Twentieth Century*. Macmillan Publishing Company, New York, 1987.
22. Denton, Edgar A. "A Chronicle: My Second Greatest Adventure (The First Being my Marriage to Mimi Elaine Tripp)" (Source: Albert P. Clark Collection of the US Air Force Academy McDermott Library)
23. Doolan, Robert H. *Personal Narrative*. Robert Henry Doolan Collection (AFC/2001/001/44182), Veterans History Project, American Folklife Center, Library of Congress, 2003.
24. *van Drogenbroek, Ben with Steve Martin. *The Camera Became My Passport Home, Stalag Luft 3, The Great Escape, The Forced March, and the Liberation at Moosburg, the memoirs of Charles Boyd Woehrle*. Stalag Luft 3 Archives of Holland and the Prisoner of War Archives of Canada, 1st edition August 2012, 2nd edition, August 2016.
25. Durand, Arthur A. *Stalag Luft III – The Secret Story*. Louisiana State University Press, 1988.

26. Eden, Paul, *The Encyclopedia of Aircraft of WWII*. Amber Books Ltd., London, 2004.
27.*Elliott, Robert D. *92nd Bomb Group (H), Fame's Favored Few.* Turner Publishing Company, Paducah KY, 1996.
28. Frances, Paul, Richard Flagg, and Graham Crisp. "Nine Thousand Miles of Concrete - A Review of Second World War Temporary Airfields in England," Airfield Research Group, Historic England, January 2016.
29. Freeman, Roger A., *The Mighty Eighth – A History of the Units, Men, and Machines of the US 8th Air Force*. Cassel & Co., London, England, 2000.
30. Freeman, Roger A. *The Mighty Eighth War Diary*. Motorbooks International, St. Paul Minnesota, 1990.
31. Freeman, Roger A. *The Mighty Eighth War Manual.* Motorbooks International, St. Paul Minnesota, 1991.
32. Friend, David. "Seeing Beyond the Stars." *LIFE Magazine,* Volume 8, Number 13, December 1985, pp. 29-42.
33. Gladwell, Malcolm. *The Bomber Mafia.* Little, Brown, and Co., Boston, 2021.
34. Grissom, Gus, *The MATS Flyer (Interview, February 1963). Interviewed by John P. Richmond, Jr. Military Air Transport Service, United States Air Force. pp. 4–7.* Retrieved June 28, 2020.
35. Hatton, Greg (392nd Bomb Group). *Stalag Luft III, American Prisoners of War in Germany*. Military Intelligence Service War Department, July 15, 1944.
36. Higbee, Arthur. "Air Force Demonstrates a Matador Missile Launch." *Oxnard Press-Courier*. California, March 24, 1956.
37. Holmstron, Carl H. *Kriegie Life*. self-published (Source: Albert P. Clark Collection of the US Air Force Academy McDermott Library)
38. Hutchison, Jane, "When Frogs Took Wing," *NASA Magazine*, Summer 1993.
39. Jablonski, Edward. *America in the Air War*. Time-Life Books, Alexandria, Virginia, 1982.
40. Jablonski, Edward. *Flying Fortress – the Illustrated Biography of the B-17s and the Men Who Flew Them*. Echo Point Books & Media,

Brattleboro, Vermont, 2014 (Corrected Edition).
41. Jones, Tom. *Sky Walking, An Astronaut's Memoir*. Harper-Collins Publishers, New York, NY, 2006.
42. Law, TSgtPeter A. "Senior Ranking Officer, A Biographical Look at Prisoner of War Leadership," Historian, 12th Flying Training Wing, December 2001 (Source: Air Force Historical Research Agency)
43. Ledden, Robert W. *Barbed Wire Interlude, a Souvenir of Krie...der Luftwaffe No, 4, Deutschland 1944*. Self-published, Alexandria VA, 1945 (Source: Air Force Historical Research Agency)
44. Maher, Daniel J. *POW Logbook, Stalag Luft III South Camp*. October 43 – May 1945 (Source: Mighty Eighth Air Force Museum Library)
45. Mansfield, Cheryl L. "If Walls Could Talk." *Missions Highlights*. NASA John F. Kennedy Space Center. (Source: www.nasa.gov website)
46. Marconi, Elaine M. "TCDT: Hands-on Launch Training." *Space Shuttle Era Missions Highlights*. NASA John F. Kennedy Space Center. (Source: www.nasa.gov website)
47.*Miller, Donald L. *Masters of the Air, America's Bomber Boys Who Fought the Air War against Nazi Germany*. Simon & Schuster, 2006.
48. Mindling, George and Robert Bolton. *U.S. Air Force Tactical Missiles, 1949-1969, The Pioneers*. Self-published, January 31, 2011.
49. Nichol, John and Tony Rennell. *The Last Escape, The Untold Story of Allied Prisoners of War in Europe 1944-45*. Viking Adult, 2003.
50. Scutts, Jerry. *Combat Legend Messerschmitt Bf 109*. Airlife Publishing, 2002.
51.*Simmons, Kenneth W. *Kriegie, Prisoner of War*. Thomas Nelson & Sons, New York, 1960 and 2019.
52. Sloan, John S. *The Route as Briefed*. Argus Press, Cleveland Ohio, 1946.
53. Smotherman, Dr. Alicia Wilkerson Smotherman. *James Franklin Smotherman, His Ancestors and Descendants*. Self-published, December 2001.
54. Smotherman, Major Benjamin F. "Missiles and Men." *TAF Review - The Twelfth Air Force Magazine*. March 1957, Vol. 3, No. 3.
55. Smotherman, Benjamin F. *PENNY comic strip*. "The Circuit, South

Compound" U.S. Edition newsletter, Number 1, April 10, 1943, Stalag Luft III, Sagan, Germany reproduced in "Prisoners of War Bulletin". American National Red Cross for the Relatives of American Prisoners of War and Civilian Internees, Vol. 2, No. 8, August 1944, Washington DC.

56. Snyder, Steve. *Shot Down, The true story of pilot Howard Snyder and the crew of the B-17 Susan Ruth.* Sea Breeze Publishing, 2015.
57. Stiles, Bert. *Serenade to the Big Bird.* Arcadia Press, 1952.
58. Webster, George. *The Savage Sky, Life and Death on a Bomber over Germany in 1944.* Stackpole Military History Series, 2007.
59. van de Weerd, Evert. "Crash No. 3 Flying Fortress." *Air war above Ede, Plane crashes and bombings in the Gelderse Vallei, 1943-1944, Chapter 4.* Barneveld Publisher BDU, Boeken 2009.
60. *Wright, Arnold A. *Behind the Wire, Stalag Luft III South Compound Records.* (Sources: Mighty Eighth Air Force Museum Library and Albert P. Clark Collection of the US Air Force Academy McDermott Library).
61. Yenne, Bille. *Hap Arnold, The General Who Invented the U.S. Air Force.* REgenery History, Washington DC, 2013.

YB-40 References

1. "Boeing-Lockheed Vega XB-40." *U.S. Air Force Fact Sheet.* National Museum of the US Air Force, June 26, 2009.
2. Bishop, Cliff T. *Fortresses of the Big Triangle First.* Elsenham, UK: East Anglia Books, 1986.
3. Doolan, Robert H. "Flying in Boeing the YB-40 Flying Fortress." *8th AF News*, Vol. 11, No. 2, September 11.
4. Dorr, Robert F. *Brilliant Mistakes: The YB-40.* September 28, 2015.
5. Doyle, David. *B-17 Flying Fortress, Vol. 2. Boeing's B-17E through B-17H in World War II.* Schiffer Publishing, Atglen PA, 2021.
6. Furniss, William E. MD. "Letter to Roger A. Freeman about YB-40s." Feb. 6, 1994. (Source: Mighty Eighth Air Force Museum Library)
7. Galland, Adolf. *The First and the Last: Germany's Fighter Force in*

WWII (Fortunes of War). Cerberus Press, South Miami, Florida, 2005.

8. Levine, Alan J. *The Strategic Bombing of Germany, 1940–1945.* Praeger, Westport, Connecticut, 1992.
9. Lloyd, Alwyn T. *B-17 Flying Fortress, Vol. 20, More Derivatives Part 3.* Detail & Scale, Inc. TAB Books, Blue Ridge Summit Pennsylvania, 1986.
10. O'Connell, Jim. "Boeing YB-40 Flying Fortress." 92nd Air Refueling Wing Historian Fairchild Air Force Base (www.fairchild.af.mil), June 21, 2013.
11. Porter, David, "The YB-40 Gunship." Military History Matters (military-history.org)

Website Resources

1. www.93bg.com (Complete History)
2. www.482nd.org (482nd Bombardment Group)
3. www.abct.org.uk (Alconbury Abbots Ripton - Airfields of Britain Conservation Trust UK)
4. *www.americanairmuseum.com
5. B17flyingfortress.de
6. www.britannica.com (General Information)
7. www.hangar6aircafe.com (Uvalde Garner Field History)
8. www.joebaugher.com (1942 USAAF Serial Numbers 42-10960 to 42-30031)
9. www.military.wiia.org
10. www.military-history.fandom.com (Operation Torch)
11. www.moosburg.org (Moosburg Online: POW camp Stalag VII A)
12. www.nationalmuseum.af.mil (National Museum of the United States Air Force)
13. www.pegasusarchive.org (POW references)
14. www.scottylive.com (The Mighty Eighth WW2 Calendar)
15. www.spaceline.org (Keith Lethbridge, Matador Fact Sheet)
16. www.stalagluft3.com (The Long March, the evacuation of Stalag Luft III, Sagan)

17. www.wendoverairbase.com (Historic Wendover Airfield Foundation - Wendover, Utah)
18. www.yumpu.com (92nd B-17 Aircraft Roster)

Other Resources

1. "Captured: American POWs in WWII" Online Course, WWII Museum and Arizona State University
2. Albert P. Clark Collection, United States Air Force Academy McDermott Library, Colorado Springs, Colorado
3. Mighty Eighth Air Force Reference Library, Pooler, GA
4. United States Air Force Historical Research Agency, Maxwell Air Force Base, Alabama

Appendix

1 BENDIX LOWER RETRACTABLE CHIN TURRET ... 400 ROUNDS PER GUN

2 SPERRY MODEL BT-44-UD-104 UPPER FORWARD TURRET ... 400 ROUNDS PER GUN

3 MARTIN MODEL 250CE4 UPPER REAR TURRET ... 400 ROUNDS PER GUN

4 BRIGGS LOWER SEMI-RETRACTABLE TURRET ... 600 ROUNDS PER GUN

5 USMC POWER BOOSTED TWIN 50 CALIBRE SIDE GUNS .. 300 ROUNDS PER GUN

6 USMC POWER BOOSTED TWIN 50 CALIBRE TAIL GUNS...550 ROUNDS PER GUN

7 RESERVE AMMUNITION FOR CHIN TURRET ... 400 ROUNDS PER GUN

8 RESERVE AMMUNITION FOR UPPER FWD. TURRET ... 400 ROUNDS PER GUN

9 RESERVE AMMUNITION FOR ALL STATIONS ... FOUR 200 ROUND BOXES AND TWENTY 150 ROUND BOXES

10 AMMUNITION BOXES FOR TAIL GUNS ... 300 ROUNDS PER GUN 250 ROUNDS PER GUN IN TRACKS

11 USMC HYDRAULIC SYSTEM FOR SIDE GUNS

12 USMC HYDRAULIC SYSTEM FOR TAIL GUNS

13 ARMOR PLATE FOR CREW PROTECTION

14 ARMOR PLATE FOR POWER PLANT PROTECTION

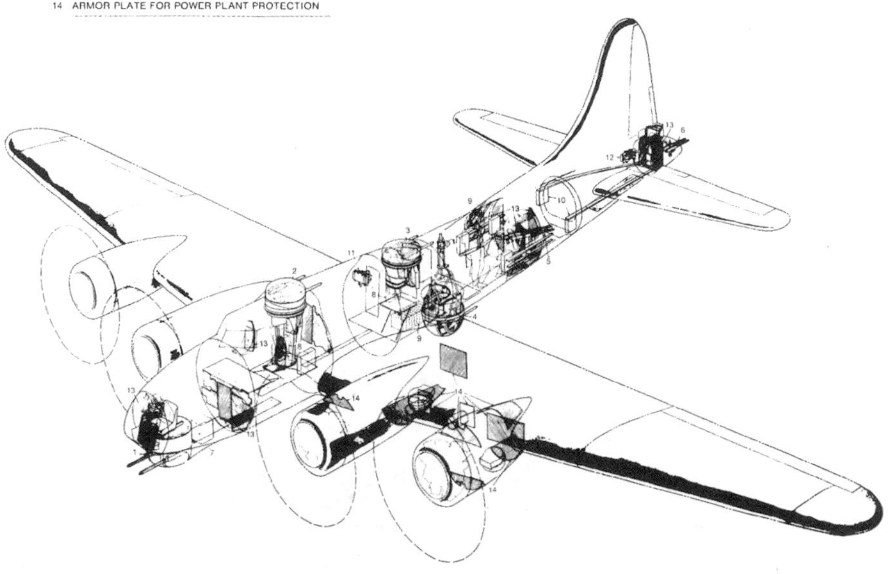

Figure I: Cutaway diagram of YB-40 (Courtesy of 92nd Air Refueling Wing Historian, Fairchild Air Force Base)

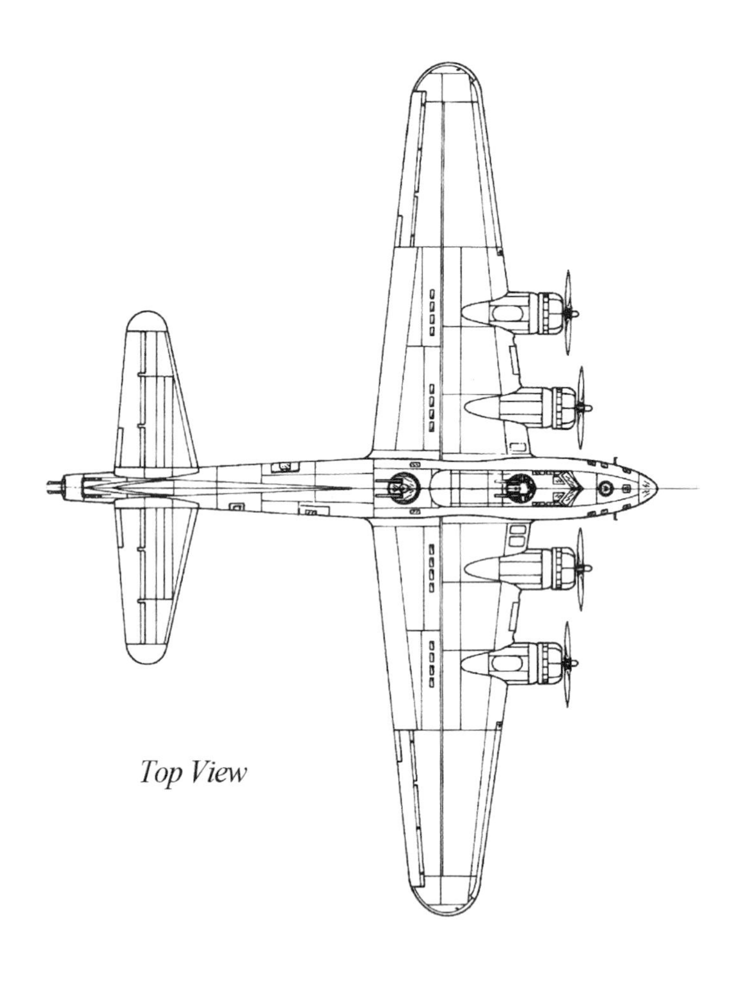

Top View

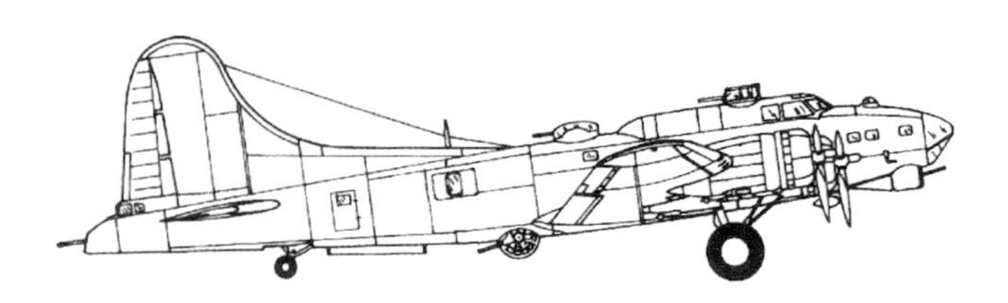

Starboard View

Figure II: Schematics of the YB-40

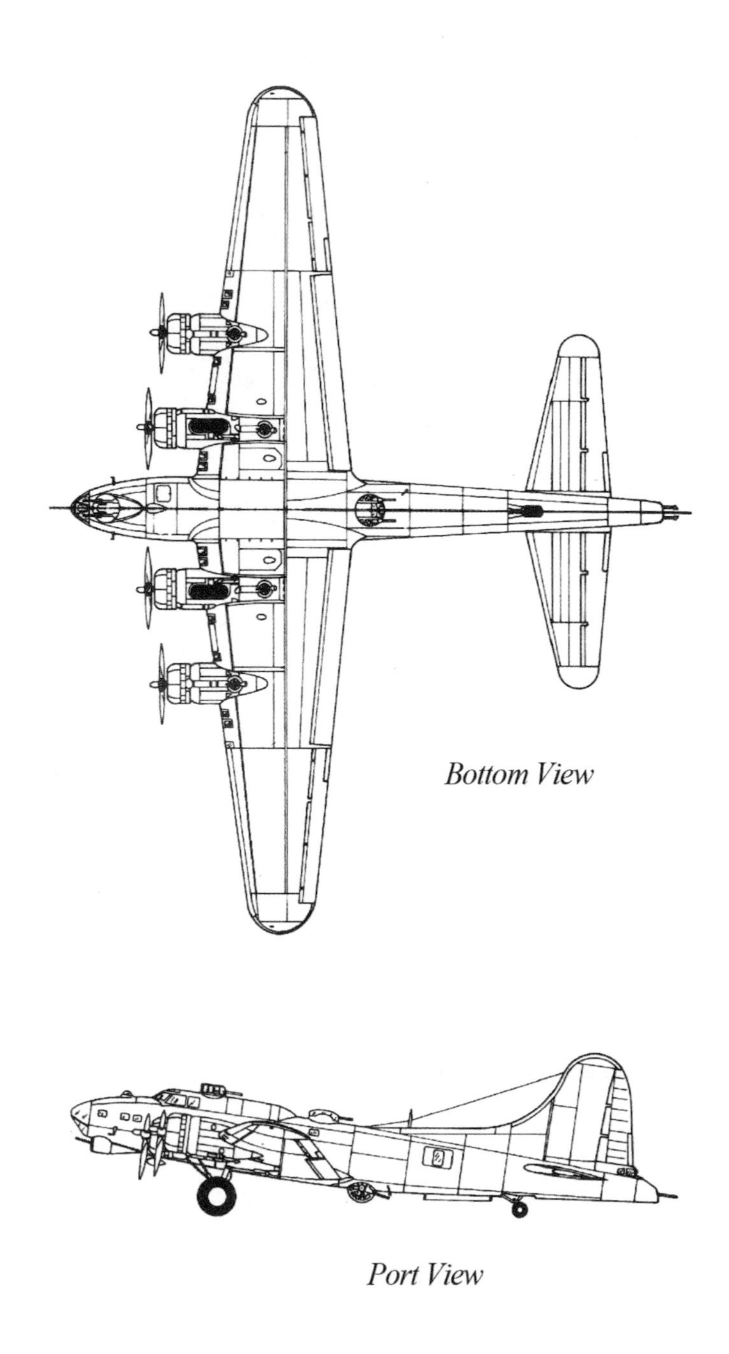

Figure II: Schematics of the YB-40 (continued)

Table I: YB-40s stationed at Alconbury, 327th BS, 92nd BG

Serial Number	Name	Livery	Date at Alconbury		
42-5732	The Mugger	UX	Crash Landed En Route	Originally B-17F-10-VE.	
42-5733	Peoria Prowler	UX-F	3-May-43	Originally B-17F-10-VE.	
42-5734	Seymour Angel	UX-D	14-May-43	Originally B-17F-10-VE.	
42-5735	Wango Wango	UX-B	4-May-43	Originally B-17F-10-VE.	Only YB-40 Lost in Combat, 22 June 43
42-5736	Tampa Tornado	UX-C	14-May-43	Originally B-17F-10-VE.	
42-5737	Dakota Demon	UX-K	11-May-43	Originally B-17F-10-VE.	
42-5738	Boston Tea Party	UX-G	4-May-43	Originally B-17F-10-VE.	
42-5739	Lufkin Ruffian	UX-J	4-May-43	Originally B-17F-10-VE.	
42-5740	Monticello	UX-E	4-May-43	Originally B-17F-10-VE.	
42-5741	Chicago	UX-H	7-May-43	Originally B-17F-10-VE.	
42-5742	Plain Dealing Express	UX-L	13-May-43	Originally B-17F-10-VE.	
42-5743	Woolaroc	UX-M	11-May-43	Originally B-17F-10-VE.	
42-5744	Dollie Madison	UX-A	11-May-43	Originally B-17F-10-VE.	Daddy Ben's Plane

YB-40's not deployed to ETO

42-5871	Originally B-17F-30-VE.	Became TB-40
42-5920	Originally B-17F-30-VE.	Became TB-40
42-5921	Originally B-17F-30-VE.	Became TB-40
42-5923	Originally B-17F-30-VE.	Became TB-40
42-5924	Originally B-17F-30-VE.	Became TB-40
42-5925	Originally B-17F-30-VE.	Became TB-40
42-5927	Originally B-17F-30-VE.	Became TB-40

TB-40s used for Training

42-5833	Originally B-17F-25-VE.	TB-40 Deployed to ETO
42-5834	Originally B-17F-25-VE.	TB-40
42-5926	Originally B-17F-35-VE.	TB-40
42-5872	Originally B-17F-30-VE.	TB

XB-40 (experimental)

41-24341	Originally the Second B-17F-1-BO	Became TB-40

Table II: YB-40 Missions

Date	Target	No. of YB-40s Dispatched	Notes
29-May-43	St. Nazaire, France	8	Mission recalled before reaching enemy coast
15-Jun-43	LeMans, France	4	Mission recalled before reaching enemy coast
22-Jun-43	Huls, Germany	11	Mission recalled before reaching enemy coast
23-Jun-43	LeMans, France	8	Mission recalled before reaching enemy coast
25-Jun-43	Oldenburg, Germany	7	Mission recalled before reaching enemy coast
26-Jun-43	Poissy, France	5	Mission recalled before reaching enemy coast
28-Jun-43	St. Nazaire, France	6	Mission recalled before reaching enemy coast
29-Jun-43	Villacoublay, France	2	Mission recalled before reaching enemy coast
4-Jul-43	Nantes, France LeMans, France	1 2	Mission recalled before reaching enemy coast
14-May-43	Villacoublay	5	
17-Jul-43	Hannover, Germany	2	Mission recalled before reaching enemy coast
24-Jul-43	Heroya, Norway	1	Mission recalled before reaching enemy coast
26-Jul-43	Hannover, Germany	2	Mission recalled before reaching enemy coast
28-Jul-43	Kassel, Germany	2	Mission recalled before reaching enemy coast
29-Jul-43	Kiel, Germany	1	Mission recalled before reaching enemy coast

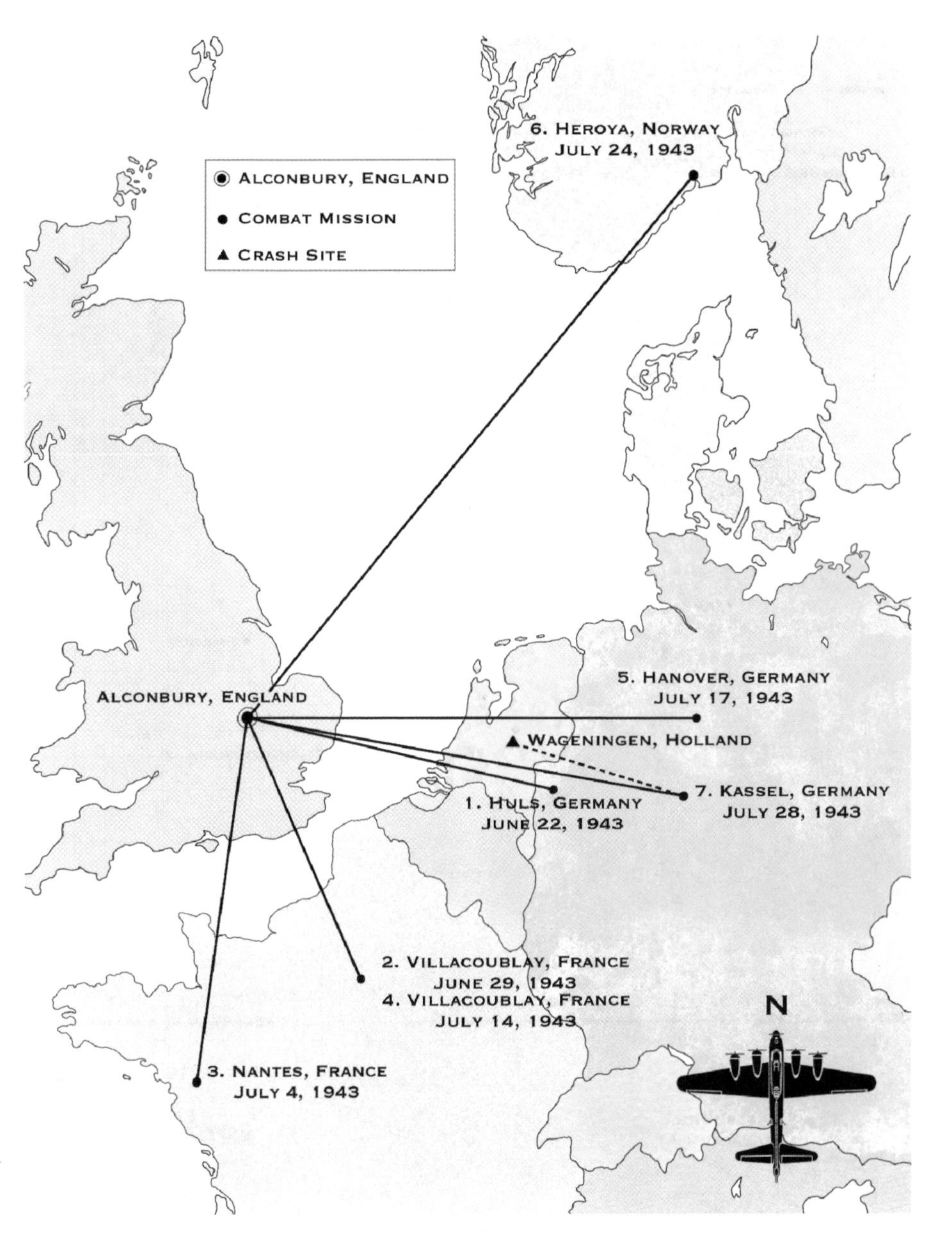

Figure III: Map of Benjamin Franklin Smotherman (BFS) combat missions. Trajectory of actual combat missions is not depicted. Target locations and relative distances to Alconbury are for reference only. (Map by Sarah Bateman Testa)

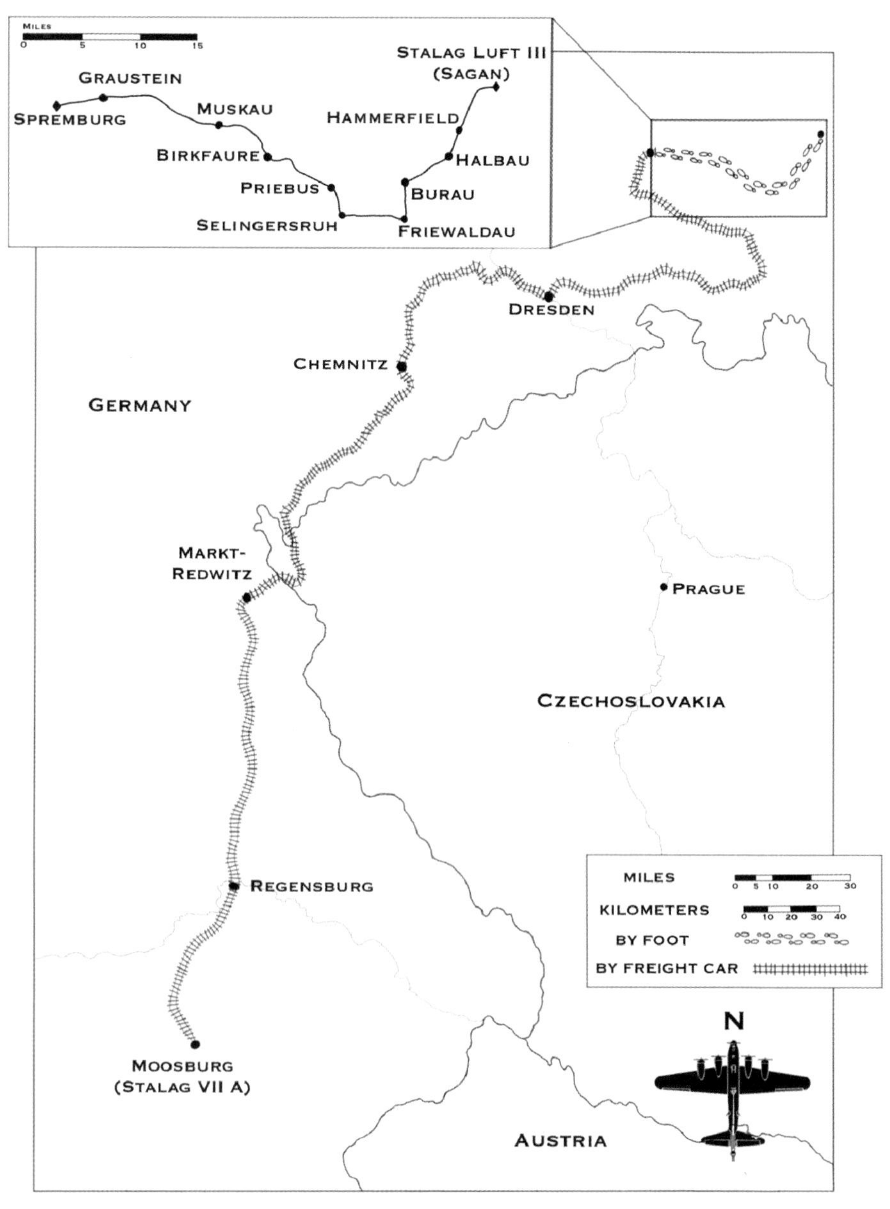

Figure IV: Route taken by Ben Smotherman on long winter march, January – February 1945. (Map by Sarah Bateman Testa)

Table III: Benjamin Franklin Smotherman (BFS) Missions

Combat Mission	Type	Date	Target	8th BC Mission	92nd BG Sortie	92nd BG Mission
1	Combat	22-Jun-43	Huls, Germany	65	12	21
-	BFS Aborted	28-Jun-43	St. Nazaire, France	69	25	15
2	Combat	29-Jun-43	Villacoublay, France	70	25	16
3	Combat	4-Jul-43	Nantes, France	71	17	28
4	Combat	14-Jul-43	Villacoublay, France	73	32	19
5	Combat	17-Jul-43	Hannover, Germany	74	33	20
6	Combat	24-Jul-43	Heroya, Norway	75	34	21
-	92nd Aborted	25-Jul-43	Kiel, Germany	76	-	35
7	Combat	28-Jul-43	Kassel, Germany	78	37	23

BC – Bomber Command
BG – Bombardment Group

Source: 92nd Bombardment Group (H) 28 Oct 1943-9 Mar 1944 Microfiche, Air Force Historical Research Agency

Table IV: Jan Davis' Space Shuttle Missions

Name	Space Shuttle	Launch Date	Landing Date	Duration	Orbits	Crew	Primary Payloads
STS-47	Endeavour	12-Sep-92	20-Sep-92	7 days, 22 hrs+	127	Hoot Gibson, CDR	Spacelab J
						Curt Brown, PLT	
						Mark Lee, PLC, MS1	
						Jay Apt, MS2	
						Jan Davis, MS3	
						Mae Jemison, MS4, SMS	
						Mamoru Mohri, PS	
STS-60	Discovery	3-Feb-94	11-Feb-94	8 days, 7 hrs+	130	Charlie Bolden, CDR	Spacehab
						Ken Reightler, PLT	Wake Shield Facility
						Jan Davis, MS1	
						Ron Sega, MS2	
						Franklin Chang-Diaz, PLC, MS3	
						Sergei Krikalev, MS4	
STS-85	Discovery	7-Aug-97	19-Aug-97	11 days. 20 hrs+	190	Curt Brown, CDR	CRISTA-SPAS
						Kent Rominger, PLT	
						Jan Davis, PLC, MS1	
						Bob Curbeam, MS2	
						Steve Robinson, MS3	
						Bjarni Tryggvason, PS